Life in the Soil

Life in the Soil

JAMES B. NARDI

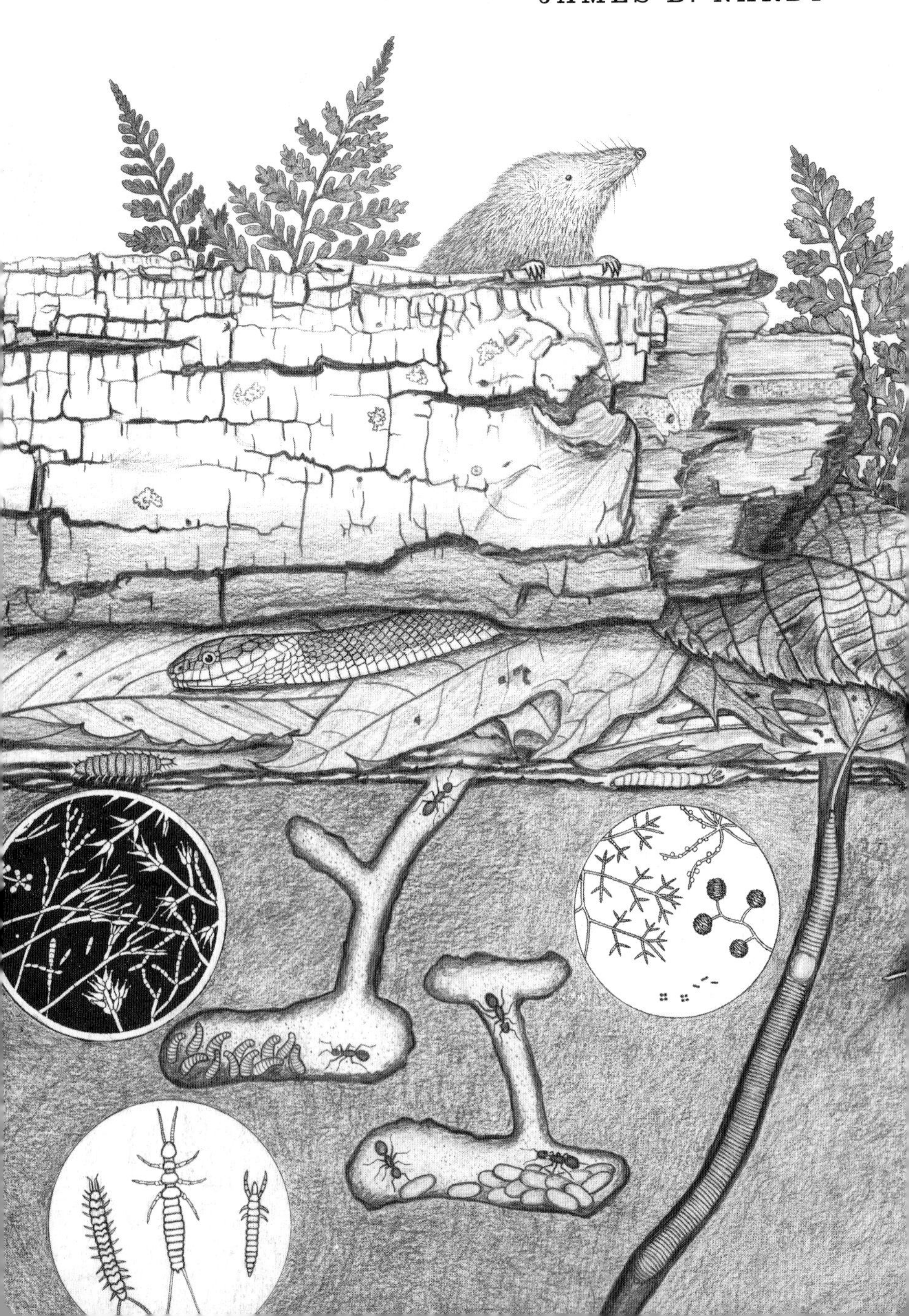

A Guide for Naturalists and Gardeners

The University of Chicago Press CHICAGO AND LONDON

JAMES B. NARDI is a biologist at the University of Illinois at Urbana-Champaign and the Illinois Natural History Survey who gardens with the help of innumerable soil creatures.

The University of Chicago Press, Chicago 60637
The University of Chicago Press, Ltd., London

Printed in the United States of America

16 15 14 13 12 11 3 4 5

ISBN-13: 978-0-226-56852-2 (paper)
ISBN-10: 0-226-56852-0 (paper)

Library of Congress Cataloging-in-Publication Data

Nardi, James B., 1948–
Life in the soil : a guide for naturalists and gardeners / James B. Nardi.
p. cm.
Includes index.
ISBN-13: 978-0-226-56852-2 (pbk. : alk. paper)
ISBN-10: 0-226-56852-0 (pbk. : alk. paper) 1. Soil biology. I. Title.
QH84.8.N36 2007
578.75'7—dc22
2006037834

∞ The paper used in this publication meets the minimum requirements of the American National Standard for Information Sciences—Permanence of Paper for Printed Library Materials, ANSI Z39.48-1992.

Contents

Plates follow page 138.

Acknowledgments

Over many years this manuscript on life in the soil grew in the fertile environment provided by a community of friends and colleagues. Some friends helped me prepare diagrams and design illustrations. Others shared their discoveries, advice, and encouragement. Each person's contribution enriched the manuscript.

My mother steadfastly nurtured my interest in nature and gardening from the time when I first encountered the world beneath my feet. Mark Bee labeled most illustrations and helped prepare many diagrams for this guide. His many talents and interests—as microscopist, musician, and artist—added to the impact of the illustrations. Joyce Scott advised me on artistic matters—proportions, textures, shading. She could see ways to improve drawings that often eluded me. Molly Scott is an accomplished artist who worked part-time at the University of Illinois Artist Service. She spent many hours scanning illustrations for this book and preparing a number of its diagrams.

Michael Jeffords contributed most of the photographs that grace this guide. As Education and Public Relations Liaison at the Illinois Natural History Survey, Michael's photographs have appeared in many publications and presentations. They speak eloquently for nature.

My Sunday morning companions in the Indiana woods, Daisy (fig.118) and Maggie, explored the soil with me and often were able to "see" with their noses creatures that I probably would not have seen with my eyes. They clearly delighted in the discoveries that they shared with me.

Friends shared their discoveries, their passions, and their knowledge of particular organisms or particular topics. As a skilled teacher for the Natural Resources Conservation Service, Ray Archuleta offered advice on preparing more effective illustrations. Don Johnson studies the sculpturing of the earth's surface by animals. He introduced me to the term "biomantle" to describe the biologically generated "epidermis" of soil. Jackie Worden introduced me to the pleasures and potentialities of using colored pencils for illustrations. Mark Sturges (the Farm at Chili Nervanos) shared his knowledge of composting as well as his enthusiasm for the creatures that are responsible for the success of organic gardening. From wherever she traveled, Pamela Sutherland would send me creatures or photographs of creatures. Susan Gabay-Laughnan furnished me with insect models from the woods near her home and from the fields where she carries out her experiments in corn genetics. On an afternoon trek through Trelease Woods, Lowell Getz introduced me to the snails and slugs of east-central Illinois. Larry Hoffman showed me how to culture algae from the soil. As chairperson of Champaign County Audubon's education committee, Beth Chato was a valuable source of information on vertebrates and their associations with soils; Ed Zaborski took me on several earthworm forays and later, at his stereomicroscope, he pointed out the features that are useful in distinguishing different species of earthworms.

Drawing from life was important in preparing the illustrations. Don Jarc found a mole in his garden that posed for the illustration in this book. In helping me search for specimens of certain arthropods, Rosanna Giordano, with her unflagging enthusiasm and determination, tracked down several species that I was unable to find in the soils of Illinois and Indiana. Finding adult and larval specimens of root maggots turned out to be more challenging than I had expected. When I contacted them for the first time, Stefan Jaronski in Montana, Amy Dreves in Oregon, and Rosanna Giordano in Vermont helped in a search that culminated with the shipment of these insects from Jay Whistlecraft in Ontario.

As enthusiastic teachers, Jennifer Anderson, Huzefa Raja, Carol Shearer, and Meredith Blackwell introduced me to the extraordinary Kingdom of the fungi. Josephine Rodriquez, Colin Favret, and Jim

Whitfield helped me find specimens of the more exotic arthropods in the collections at the Illinois Natural History Survey and the University of Illinois. Gary Olsen brought me up to date on the classification of Eubacteria and Archaebacteria. In late winter, long before I could find any gastropods for examination of their radulae, Betty Ujhelyi brought me some from her backyard.

My colleagues who study insects and their relatives often helped with finding and identification of specimens. Don Webb (flies and scorpionflies), David Voegtlin (aphids and phylloxerans), and Felipe Soto (springtails and diplurans) were always eager to discuss their favorite arthropods.

For many years Heather (plate 66) and Francis Young shared their composting expertise with our east-central Illinois community. A few years ago they returned to their native New Zealand, where their composting and organic gardening continue unabated.

During preparation of the complete manuscript for submission to the editorial office, Jerald Kimble formatted the text with great care to meet the publisher's specifications.

As editors for this guide, Christie Henry and Michael Koplow at the University of Chicago Press were constant sources of good suggestions. Their imagination, advice, and enthusiasm were unflagging. The artistic blending of illustrations and text reflects the care and talent of the designer for the guide, Matt Avery.

How to Use This Book

This book is divided into three parts. Part 1, "The Marriage of the Mineral World and the Organic World," explains how organisms are specially adapted for life underground and examines their relationship with the chemical and physical properties of soils. Part 2, "Members of the Soil Community," considers the contributions of particular groups of creatures to soil communities. This second part is organized in three sections: microbes, invertebrates, and vertebrates. Part 3, "Working in Partnership with Creatures of the Soil," discusses ways to enrich the soil of our farms as well as gardens, including how to start a backyard compost pile.

The book concludes with a discussion of ways to collect and observe creatures of the soil, followed by a glossary of scientific terms that are used at least once in the text and a list of books for additional reading.

FACT BOXES

A fact box accompanies each entry in Part 2. Each fact box is a quick reference that contains information about the common name for the group of organisms, its classification, place in the food web, its impact on gardens, its size, and number of species in the group worldwide. Fact boxes in the vertebrates section also contain information on life spans and gestation periods. Other text boxes appear throughout the book.

Common Name

The title of the fact box gives the common name(s) for the group of organisms.

Classification

In order to keep track of relationships among organisms, creatures are grouped and named according to the features they share. The name or classification of each organism proceeds in a definite sequence. Groups of organisms are arranged in a definite order, or hierarchy, from the smallest number of groups (domains and kingdoms) to the largest number of groups (genera and species), from the most general at the topmost level to the most specific at the bottom of the hierarchy.

The most general are the domains, and there are three domains that include all organisms. Archaea (also known as Archaebacteria) and Bacteria (also known as Eubacteria) are the two domains of prokaryotic (*pro* = before; *kary* = nucleus) organisms, whose cells lack true membrane-bound nuclei. Collectively, the prokaryotic organisms in these two domains are known as bacteria. The third domain, Eukarya, includes all eukaryotic (*eu* = true, good; *kary* = nucleus) organisms, whose cells have true nuclei with delimiting membranes.

Each of the prokaryotic domains is also considered a kingdom: Archaebacteria and Eubacteria. The domain Eukarya is divided into several kingdoms; but the number of these kingdoms, the grouping of certain organisms into these kingdoms, and the naming of these kingdoms has changed several times in recent years. One widely recognized organization in the domain Eukarya places organisms into five kingdoms: Protozoa, Chromista, Fungus, Animal, Plant (fig. 1). The Fungus, Animal, and Plant kingdoms are universally recognized by biologists, but the placement of eukaryotic organisms that are not members of these three kingdoms is often contested. Microbes are found in all domains and represent creatures that cannot be readily seen without a microscope. The prokaryotic microbes are represented by the bacteria; the eukaryotic microbes are generally referred to as protists, fungi, and algae. Vertebrates and invertebrates are the two divisions of the Animal kingdom.

The most specific groups in the classification of organisms are species, and the number of known species on Earth is currently well over a million. The levels in the hierarchy of classification are

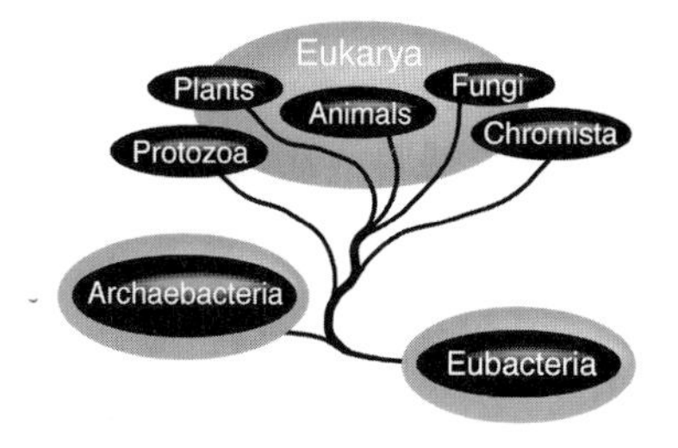

1. The tree of life has three domains—Eukarya, Archaebacteria, Eubacteria—and seven kingdoms—Archaebacteria, Eubacteria, Protozoa, Chromista, Plants, Fungi, and Animals.

Domain
 Kingdom
  Phylum
   Class
    Subclass
     Order
      Suborder
       Superfamily
        Family
         Subfamily
          Genus
           Species

Each group is given a name often derived from Latin or Greek that tells something about the group. For example, slow-moving organisms known as tardigrades are classified in the phylum Tardigrada, which is Latin for "slow stepper." Biologists name each organism according to one or more of these levels of classification.

Most groups of organisms featured in this guide include creatures that are universally encountered in soils. These groups each consist of just a few to many thousands of species; the common features of each group—whether the group is a genus, family, order, class, phylum, or kingdom—are described, and at least one representative of each group is illustrated. In some soils, these organisms can be found in great numbers. However, a few exceptional groups of organisms—such as onychophorans, ricinuleids, or caecilians—are found not universally but only in certain climates. To add to a comprehensive coverage of life in Earth's soils and to emphasize the great diversity of this life, however, those small groups with unusual and distinctive characteristics have also been included in this guide.

Place in Food Web

The food web is a network of organisms within which energy and nutrients—the substances that all organisms in every kingdom need for survival—are exchanged. Ultimately each organism dies and returns to the soil; but during its life, it plays one or more of the following roles in the food web:

Algal eaters—organisms that eat algae.
Bacterial partners of plants—bacteria that live within plant roots and form a mutually beneficial relationship with plants.
Coprophages—organisms that feed on droppings and dung.
Decomposers—organisms that break down the remains or waste products of other organisms.
Detritivores—organisms that feed on dead plant and animal matter.
Diggers—organisms that facilitate the circulation of nutrients between layers of soil and help stimulate growth of plants.
Fungal partners of plants—fungi that form mutually beneficial relationships with plant roots.
Fungivores—organisms that feed on fungi.
Herbivores—organisms that feed on plants.
Parasites—organisms that live in or on other living organisms (hosts) and obtain nutrients from their hosts, usually without killing them.
Predators—organisms that obtain nutrients from other living organisms (prey) but do not live in or on their prey.
Producers—organisms that produce their own nutrients from only air, water, minerals, and energy.
Scavengers—organisms that feed on dead plant or animal matter.

Impact on Gardens

Ultimately all members of the food web derive their energy and nutrients from the energy of our sun. Through photosynthesis, green plants, algae, and certain bacteria capture the sun's energy to produce nutrients from only the raw materials of air, water, and minerals. Plants are primary producers whose photosynthetic products represent the initial conversion of energy to nutrients within the food web, but they also depend on other members of the food web for their sur-

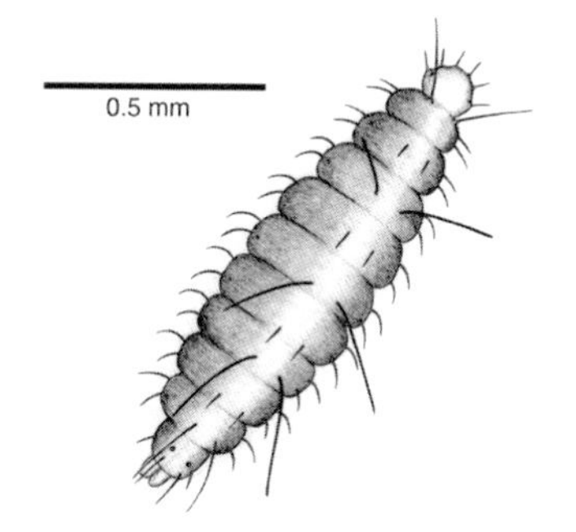

2. So many species of creatures live in the soil that many, such as this fly larva, have never been named or studied.

vival and nutrition. In our gardens, the interactions of members of the food web can have profound impacts on the well-being of vegetables and fruits. Those members of the food web that have a positive or negative influence on the health of gardens are respectively designated *allies* or *adversaries*. Some diverse groups of organisms, such as click beetles, with about 9,000 species, include both allies and adversaries of gardens. Some of the larger creatures listed in this guide are rarely or never found in the limited habitat of the garden. They are absent from gardens but are found in forests, deserts, or prairies.

Size

Sizes are given in metric units—micrometers (μm), millimeters (mm), and centimeters (cm). There are a thousand micrometers in each millimeter and ten millimeters in each centimeter. Microbes, measured mostly in micrometers, are so small that they will go unnoticed unless they are viewed under a magnifying lens or a microscope. Although the sizes of most soil creatures are specified in length, the size of an organism that has the form of a filament, such as a fungus or an alga, is given as the diameter of the filament. The size of a bird or a mammal includes the length of its head and body without its tail.

Number of Species

The number of species found worldwide is often an estimate because new species continue to be discovered, especially in groups of organisms representing many of the smaller creatures. For these groups, the number of species described in the scientific literature often represents only a small fraction of the species that actually exist (fig. 2). For the larger vertebrates, almost all species have been discovered. The number of vertebrate species, however, may change slightly from year to year. This fluctuation arises because biologists often debate whether populations that are considered one species should really be considered more than one species or whether populations that are placed in more than one species should be grouped as a single species.

Preface

Minerals from the soil nourish all creatures during their lives; and unto minerals of the soil all creatures ultimately return at their deaths. A soil is not fertile and complete until creatures occupy it and contribute their organic portion to the mineral portion of the soil. Soils represent a marriage of the mineral world and the organic world; and a definition of soil that covers these attributes very well is "a dynamic natural medium in which plants grow made up of both mineral and organic materials as well as living forms."

Whatever takes place each day in this world beneath our feet has wide-ranging influences on some of the great issues of our time—pollution, nutrition and health, global warming, and preservation of biodiversity. As creatures of the world aboveground, however, we humans have difficulty relating to this dark, hidden world. The complexity of the physical, chemical, and biological interactions taking place belowground, as well as their influence on life aboveground, has long been appreciated by practitioners of sustainable agriculture and organic farming; but now, a broader appreciation and understanding of life in this alien world has emerged from the unparalleled scientific advances of recent years. A special 2004 issue of *Science* entitled "Soils—The Final Frontier" reflects the new attention focused on this rich, varied world beneath our feet that sustains the rich biodiversity of life aboveground.

A staggering number of individual organisms as well as species, representing all kingdoms—plants, animals, fungi, protozoa, bacteria—live in the soil. The creatures that are discussed in this book represent the major groups of these organisms. Many groups of organisms never leave the soil; others live only part of their lives or part of each day in the soil. Some may be extremely abundant in some places but absent in others. Many live in dung, decaying plants, or dead animals that will eventually return to soil. The variety of soil types and the health of soils are reflected in the abundance and diversity of their living inhabitants that either directly or indirectly contribute to a soil's fertility.

Countless small, retiring inhabitants carry out their indispensable business in the soil. Although we tend to underestimate the abilities and the significance of such diminutive organisms, they happen to be the recyclers that can produce that key organic ingredient of healthy soil known as *humus,* the substance that gives soil its rich, dark color and its crumbly structure. After these creatures of the soil have thoroughly digested plant and animal debris that falls to the ground, humus is the substance that remains in their droppings. Humus is the substance that has the ability to latch onto nutrients and keep them within easy reach of plant roots as these nutrients move back and forth from the living world to the nonliving world. There is a very good reason why organic farmers add manure and compost to soil rather than commercial fertilizers. Compost and manure come with a good supply of nutrients and with their own populations of soil creatures that convert the manure and compost to humus. Humus not only holds soil nutrients and water within reach of plant roots and but also gives soil a spongy, crumbly structure that makes it particularly hospitable to creatures of the soil community. What imparts fertility to a soil is the humus that is generated as a communal effort by the community of soil creatures.

A backyard compost heap is a good place to see humus formation in action. Compost represents a microcosm of the soil surface and offers an alluring view of what transpires every day in the leaf litter of a forest or the plant debris of a prairie. The speed and thoroughness with which plant debris breaks down in a compost heap is a testament to the efficiency of the creatures who convert plant matter to humus. Stirring up a compost heap as well as poking around in the leaf litter of a forest or prairie turns up a great number and variety of decomposers that are responsible for recycling and humus formation.

These decomposers, scavengers, saprophages, recyclers—whatever you wish to call them—of dead plant and animal matter are the most abundant creatures of soils and compost heaps. Another group of soil animals, ranging in size from minuscule arthropods to corpulent rodents, have virtues as movers of earth that are all too often overlooked. Diggers keep soil in circulation, mixing lower mineral layers with the upper organic layers and enriching the soil in ways that fertilizers never manage to do.

Bacteria, however, are the only soil organisms that not only recycle the remains of plants and animals but that also liberate all the essential elements from soil particles in forms that living plants need for their survival. All life of the soil as well as all other forms of life on earth ultimately depend on these bacteria and the plants that they support.

A variety of predators—fungi, protozoa, mites, centipedes, beetles, birds, and snakes—constantly keep populations of the more abundant decomposers and diggers in check. In a healthy soil, with its mix of decomposers, diggers, and predators, a harmony pervades both the mineral world and the organic world. The creatures of the soil community depend on each other to maintain the balance between processes of growth and processes of decay.

Healthy soil, good nutrition, and good health go hand in hand; and those creatures that contribute to the health of soils are obviously of the utmost importance for the well-being of the planet. They are the ones that steadfastly keep the world running and include creatures as strange and as bizarre as any that a science fiction movie can offer. Yet we really know very little about so many of them. Leonardo da Vinci's observation that "we know more about the movement of celestial bodies than about the soil underfoot" is sadly as true today as it was in the fifteenth century. The intent of this book is to assure the reader that new discoveries and new surprises await those who stop to look a little closer at these creatures in "the soil underfoot," the unsung heroes that give the gift of good earth.

PART ONE

Marriage of the Mineral World and the Organic World

A. INTRODUCTION

Every rock has what it takes to be part of a soil someday. Even the hardest of rocks will eventually succumb to the unremitting action of weather and plants. In the early days of the earth there was neither soil nor living creatures—only rocks, water, the wind, and the sun. But these were ingredients enough for the making of the earth's first mineral soils. Four major factors worked together to form the first mineral soils: the weathering of rocks from which soil originated (*parent materials*) was influenced by the slope of the land (*topography*) where rocks were exposed to wind, rain, and sun (*climate*) as well as by the length of *time* that rocks were exposed to weathering. Soil began to form slowly, imperceptibly as large rocks changed into small rocks, and small rocks changed into even smaller rocks (plate 1).

The ultimate result of the weathering of most rocks is the formation of sand, silt, and clay—the three main mineral particles that make up all soils and that give each soil its distinctive texture. New soils and their mineral particles of sand, silt, and clay are born as rocks weather and disappear. With the appearance of these three types of mineral particles, soils begin to take on particular characteristics.

The particular combination of mineral particles found in a soil determines that soil's *texture. Sand* particles that range in diameter from 0.05 mm to 2 mm impart a coarse texture to sandy soil. Minuscule particles of *clay,* on the other hand, which are smaller than 0.002 mm in diameter impart a sticky texture to clay soil. *Silt* particles, whose di-

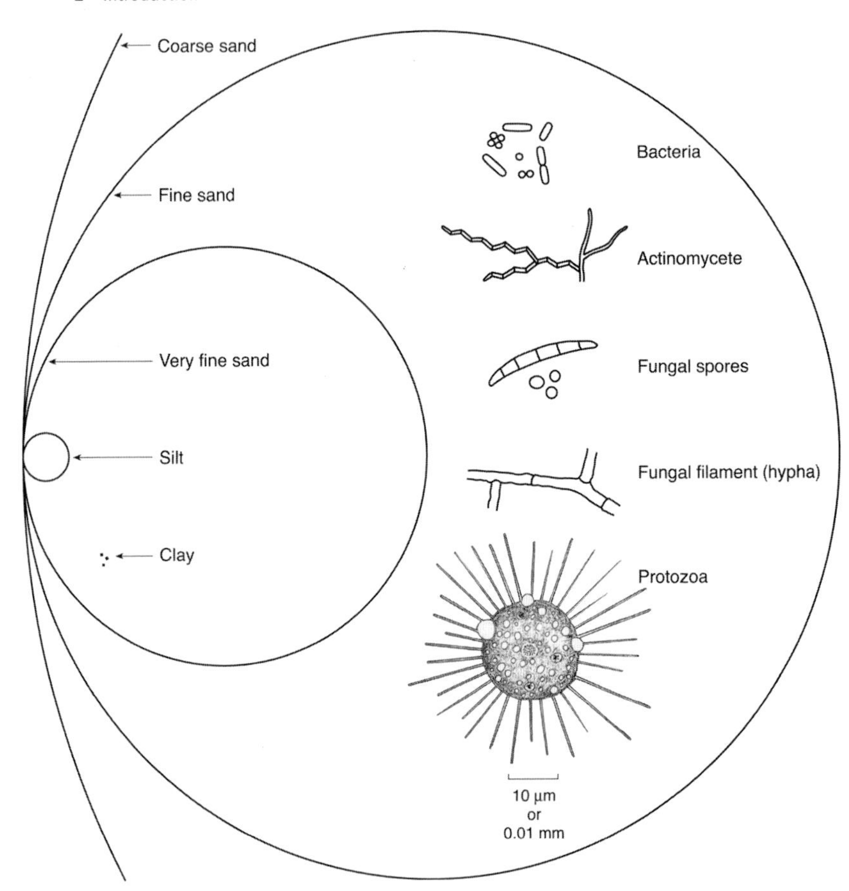

3. A world of soil creatures can live on a grain of sand. Here the sizes of soil particles are compared with those of soil microbes.

ameters are smaller than those for sand particles but larger than those for clay particles, give silt a silky or powdery texture (fig. 3).

A soil in which the stickiness of clay particles, the silkiness of silt particles, and the coarseness of sand particles contribute almost equally to the soil's texture is known as *loam* soil. All other soil textures with names like clay loam, sandy loam, silty clay, or sandy clay represent combinations of soil particles with textures lying somewhere between any two of the three major types of soil texture: sand, silt, and clay (fig. 4).

The soil-forming forces are as effective now as they were when the earth was much younger. Soil is being born today just as it has been

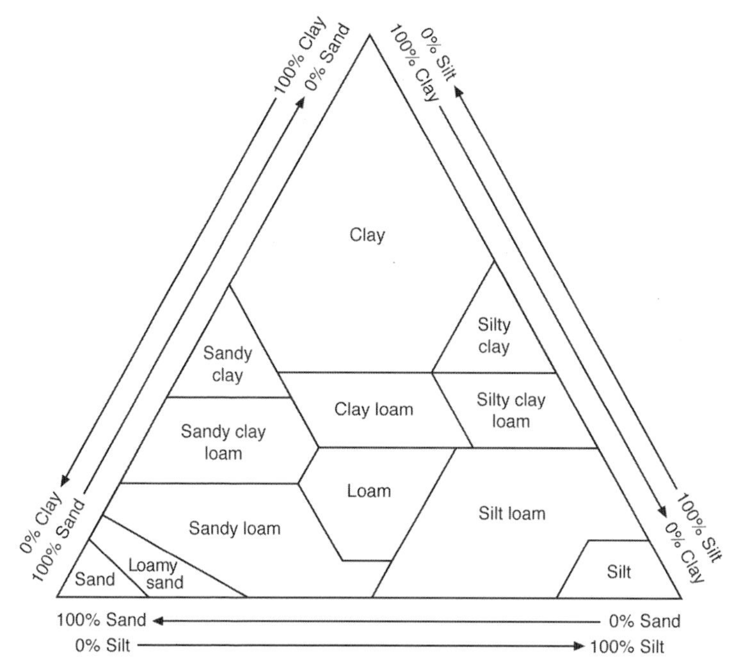

4. Textures of soils arise from different mixes of mineral particles. At a particular corner of the triangle 100 percent of the particles are either sand, clay, or silt. The percentage of this particle then diminishes gradually to 0 percent at each of the other two corners. The various soil textures are classified according to their particular percentage of sand, silt, and clay.

since the early years of the earth. But the birth of soil is a slow and labored process. By one liberal estimate, just an inch of soil takes on the order of 500 years to form; and by a more conservative estimate, one inch of soil forms every 1,000 years.

A complete soil represents the marriage of the mineral or inorganic world with the organic world. It is a good marriage; and as in all good marriages, the two partners work together in harmony. Each partner's attributes are often enhanced in the other's company. Minerals come from the breakdown of rocks; organic matter arises from the decay of animals and plants. Minerals provide many essential elements for plant life, but so does organic matter. The marriage of the mineral world and the organic world is a marriage that improves the longer the two partners work together. In addition to the four major factors (*climate, topography, parent materials, time*) that contribute to the formation of mineral soils, *living organisms* represent a fifth and final factor that not only controls but also completes the formation of soil. All these factors work together to create the great diversity of the earth's soils, which differ so much from place to place.

Mineral soils form from rocks, air, and water. These soils, however, are missing one final element that is essential for the survival of all creatures. This element is nitrogen. Ironically, soil is surrounded by this element. Even though roughly three-fourths of the air we breathe is nitrogen in the form of dinitrogen gas (N_2) and 34,500 tons of this gas lie over every acre of land, very few of our fellow creatures can use nitrogen as it exists in the air—nor can we. Animals depend on plants and other animals as sources for their nitrogen; and the only forms of nitrogen that green plants can use are ammonia (formed by combining nitrogen with hydrogen) and nitrates (formed by combining nitrogen with oxygen). In the early years of the earth, lightning was the only agent that could transform nitrogen of the air into nitrogen of nitrates. As life began to appear on earth, a few specialized bacteria, blue-green algae or cyanobacteria (*cyano* = dark blue; *bacteria* = rod), and certain actinomycetes that live in the soil began to take on the task of converting dinitrogen gas to ammonia, a process known as nitrogen fixation.

Rain, wind, sun, and ice all help convert rock to soil, but it is the living creatures that clearly make the soil hospitable for other living forms. Only hardy and intrepid pioneers can move in and eke out a meager existence on the barren and nutrient-poor stretches of uninhabited soil and rock. New pioneers appear and new nutrients are added as old pioneers pass away. As early generations of microbes, plants, and animals die and decay, later generations of creatures take up residence among their remains. Once the first pioneers are established on a rocky and barren soil, they begin to create soil particles from rock and from the remains of the creatures that preceded them. The slow and often imperceptible breakdown of rocks and the buildup of soil by the weather now proceed a little faster with help from plants, their roots, and their various partners.

Alliances between organisms have been very important, if not essential, in colonizing soils for the first time. By sharing their talents and working together, the partners survive where either one alone would have perished. Four very successful partnerships are (1) the algae and fungi that form lichens, (2) the plants and bacteria that form nitrogen-fixing root nodules, (3) the fungi and plants that form mycorrhizae, and (4) the diggers and decomposers that circulate and recycle the soil nutrients that plants need for survival.

Lichens and algae are usually the first obvious signs of life on barren, uninhabited surfaces of the earth. The rugged, tenacious lichen is half alga and half fungus and combines the best attributes of each partner. The algal partner not only captures the energy of sunlight to produce sugars and oxygen but can also fix nitrogen from the air while the fungal partner provides water and essential elements for survival and growth. Lichens, of all creatures, are especially gifted when it comes to breaking down rocks. First of all, lichens are extremely abundant and occupy 8 percent of the earth's land surface. Second, a lichen can live an awfully long time in one place on a rock—a few hundred, even a few thousand years (plate 2). A lichen can afford to be unhurried and persistent in its affairs. Third, lichens can penetrate and, after many generations, ultimately eat their way into their rocky homes. What lichens have that other organisms do not is the ability to produce a variety of acids, many of which are found nowhere else in the kingdoms of life. The names of some of the acids are taken from the scientific names given to the lichens that produce them: usnic acid of *Usnea* lichens, lobaric acid of *Lobaria* lichens, gyrophoric acid of *Gyrophora* lichens, and evernic acid from *Evernia* lichens.

Mosses too can gain footholds on bare rock. They grip the surfaces of rocks with tiny rootlets or rhizoids. These green pioneers manage to obtain enough water and elements from surfaces of rocks to sustain them as generation after generation of mosses claim the bare rock for their home. Mosses, unlike lichens, are not known to produce special acids that can digest rocks, but no one has really looked into this matter very carefully. Even if mosses only occupy space on rocks and do not send their rootlets into the rock, they still shelter bacteria, fungi, tardigrades, and other tiny organisms among their rootlets that may help transform rock to soil. Certainly some fungi can chew away at rocks. The acids secreted by the thin, delicate tips of fungal strands or hyphae allow them to perforate even the most substantial rocks.

Atop barren rocks and the sands of deserts one of the first signs of life is the formation of a thin, fragile crust, known among ecologists as a cryptobiotic (*crypto* = hidden; *bios* = life) soil, that harbors bacteria, lichens, algae, protozoa, some fungi, and some mosses (plate 3). These pioneers settle down in these sterile environments, adding nitrogen, storing water, stopping erosion, and just generally enriching the soil by intertwining their living filaments with the lifeless grains of sand and the solid crystals of rocks. But none of these pioneers has roots that can delve deep beneath the surface of the new soil. Hardy plants are drawn to these fragile sites, extending their roots and soon transforming the sites to green oases where other creatures can gain a foothold in the young soil. These plants with roots get a great deal of help along the way from bacteria, from fungi, and from animals of the soil that form some remarkable partnerships with their green allies.

B. HOW SOIL FORMS FROM ROCKS AND WEATHER

Sun, ice, water, and wind are all constantly changing the face of the earth. Where once a field of boulders covered the landscape, grasses and forbs may spread their roots into a rich, dark soil. Where once the granite roots of a mountain anchored it to the earth, massive trees now spread their own roots and branches.

As the sun beats down on a rock, its surface heats up, while just beneath the rock's surface, where the heat does not penetrate, the temperature is many degrees cooler. The rock's surface expands during the heat of the day and contracts during the cool of the night, while the well-insulated core of the rock remains unperturbed by the rising and falling temperatures. After many cycles of expansion and contraction, the outer layers of the rock begin to flake off, and a first step in the conversion of rock to soil has begun. The deserts of the world are the best places to observe this gradual wearing down of rocks by the intense heat of the midday sun and the cold of the desert night.

As water freezes in cracks of rocks, it expands to make more and wider cracks. Every time water freezes, it expands almost 10 percent in size. Freezing water can expand enough to completely split a rock. By acting as a wedge in crevices and cracks, ice contributes to rock breaking and the early stages of soil formation.

Wind works best as a former of soil when it works together with dust and sand, blowing, blasting, and scraping whatever rocks lie along its path. The rocks are eventually worn smoother and smaller as a few more mineral particles are sloughed off by each gust of wind.

Over centuries and over millennia, rain, snow, and flowing water wear away at rocks by the force and friction of their movements. Look at any rock canyon or rocky shoreline to see how water has sculpted and shrunk its surface, carrying off tiny particles of rock that may some day form part of a rich soil, sometimes far from the rocks where they first arose.

Raindrops help convert rock to soil by two different routes. The force of raindrops falling from the sky physically wears away at rock surfaces to form particles of sand and silt. Raindrops also mix with the carbon dioxide in the air to form a weak acid called carbonic acid—the acid that is responsible for the fizz of carbonated beverages like champagne and sodas. Carbonic acid is very effective at corroding surfaces of rocks like limestone.

Chemical corrosion of rocks by rain is a process of interaction and exchange of chemical elements that can be expressed in the simple, straightforward shorthand of a chemical equation. The *elements* in the equation are chemicals that cannot be broken down to other chemicals with different properties, but these elements can join to form *compounds* that are made up of more than one element. For example, hydrogen and oxygen, which are elements, combine to form water, a compound. Carbon, in turn, can combine with hydrogen and oxygen to form simple sugar compounds as well as long chains of sugars like cellulose. Each element in a chemical reaction is represented by a capital letter or a capital letter and a small letter (for example, C = carbon; Ca = calcium). Some elements and some compounds have either a negative charge or a positive charge; such charged particles are called *ions.* When nitrogen (N) and oxygen (O) join they often form negatively charged nitrates (NO_3^-), but when nitrogen and hydrogen join they form positively charged ammonium (NH_4^+). In every chemical equation, the number and the types of elements as well as the number and types of charges on the left side of the equation (the reacting chemicals) must equal the number and types of elements and charges on the right side of the equation (the chemical products). Therefore another and a quicker way to say that rain reacts with carbon dioxide to form carbonic acid is to write:

$$\underset{\text{rain}}{H_2O} + \underset{\text{carbon dioxide}}{CO_2} \rightarrow \underset{\text{carbonic acid}}{H_2CO_3}$$

The carbonic acid of raindrops is corrosive and chemically transforms rocks by removing certain elements from them. What actually happens is that the positively charged hydrogen in carbonic acid replaces other positively charged elements in the rocks and over time dissolves rocks like limestone and granite. Each rock is made up of one or more minerals, and each mineral consists of several elements. Caves are formed and calcium ions are liberated by the corrosive action of carbonic acid on limestone, otherwise known as the compound calcium carbonate ($CaCO_3$).

$$H_2CO_3 + CaCO_3 \rightarrow Ca^{+2} + 2OH^- + 2CO_2$$

A variety of clays are also formed as rocks like granite and schist encounter carbonic acid. Like all acids, carbonic acid releases positively charged hydrogen ions that continually exchange places with other positively charged elements like aluminum, magnesium, iron, sodium, and potassium that make up many of the compounds found in rocks and soils. The more rain that falls, the more carbonic acid forms, and the more hydrogen ions are added to the soil. These hydrogen ions by their sheer numbers continually displace other positive ions from rocks and contribute to the slow and steady conversion of rocks to particles of clay, sand, and silt.

The most abundant minerals of many rocks such as granite are known as feldspars. Feldspars come in a variety of forms. What they all have in common are the elements aluminum (Al), oxygen (O), and silicon (Si). All feldspars contain either one or two additional elements. These can be potassium (K), calcium (Ca), sodium (Na), or barium (Ba), with potassium being the most common element of the four. When feldspars encounter carbonic acid, the two of them react and leave behind several compounds.

$$\underset{\text{feldspar}}{2K(AlSi_3O_8)} + \underset{\text{water}}{H_2O} + \underset{\text{carbonic acid}}{H_2CO_3} \rightarrow \underset{\text{potassium carbonate}}{K_2CO_3} + \underset{\text{clay}}{Al_2Si_2O_5(OH)_4} + \underset{\text{sand/silt}}{4SiO_2}$$

One of the products of the reaction is formed when carbonate anions (negative ions) join with potassium cations (positive ions). Large particles of silicon dioxide, better known as sand, as well as smaller particles of silicon dioxide, better known as silt, result as different elements change partners during the reaction. The particles of clay are a mixture of aluminum, silicon, oxygen, and water.

C. PLANT ROOTS AND THEIR BACTERIAL PARTNERS

As soil begins to appear on the surface of sand or rock, every now and then a seed lands, germinates, and extends its newly formed roots into the young soil, testing the new environment in which it has happened to alight. Even though these soils have most of the mineral nutrients that a plant needs to get a start, they are still deficient in that one essential nutrient—nitrogen.

Because practically all of a soil's nitrogen is stored in organic rather than mineral matter, rocks and soils that contain no living things also have no organic matter and little or no nitrogen. However, some of the plants that pioneer these primordial soils have the ability to establish themselves—root, stem, and shoot—on nitrogen-poor soil where no other plants besides lichens, algae, and mosses have managed to gain a foothold.

The roots of these pioneering plants have established special partnerships with certain nitrogen-fixing bacteria called rhizobia (*rhizo* = root; *bio* = life) and other filamentous bacteria called actinomycetes. The roots accommodate these bacteria in special knots or nodules. Since plants with bacterial nodules carry their own supply of nitrogen compounds, they are able to grow where most other plants would starve. Each root nodule shelters millions of bacteria and provides a perfect environment for bacteria that fix nitrogen by converting dinitrogen gas of the air to the compound of nitrogen and hydrogen called ammonia. Although neither bacteria nor plants can use nitrogen in the form of dinitrogen gas, both can use nitrogen in the form of ammonia.

Each root is covered with very fine hairs that project at right angles to its surface, giving the root a distinctly fuzzy appearance (fig. 5, plate 4). All these root hairs, even though each is only a fraction of a millimeter in diameter, vastly expand the area of a root that comes in contact with soil as well as all the minerals and creatures inhabiting it. The root hairs of each nodule-bearing plant secrete substances that attract bacteria with the ability to fix nitrogen.

However, not just any root hairs and any bacteria can get together. Each plant requires a special type of rhizobium or actinomycete; and only when substances on the root hairs complement substances on the bacteria do the plant and the bacteria strike up an intimate relationship. Those soil bacteria that encounter compatible root hairs latch on to the hairs and spread into the root; there they stimulate

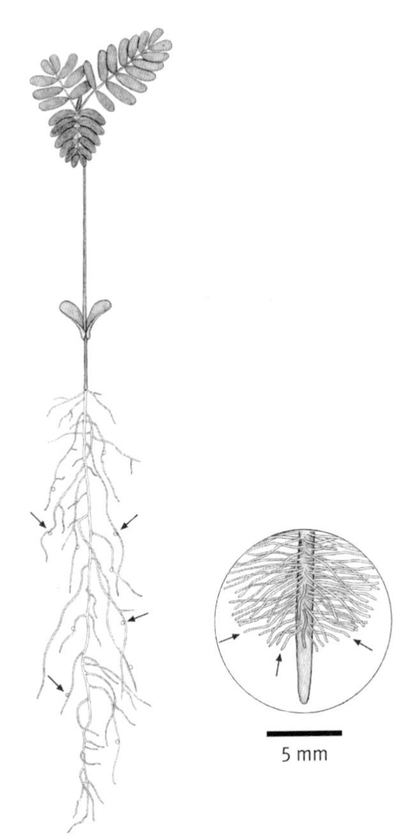

5. Each nodule (arrows) on the roots of a partridge pea contains millions of bacteria. These microbes enter the root through the fine root hairs (arrows in inset) that cover the tip of every root.

root cells to divide and form a nodule of cells (fig. 6). These bacteria have an enzyme that converts dinitrogen from the air to ammonia. The plant roots in turn supply the energy that the enzyme needs to carry out nitrogen fixation. The ammonia produced in the root nodules nourishes both plant and bacteria.

Their teamwork allows them to settle in soils that plants without nodules would never consider proper environments to sink their roots into. But after they have been settled on an inhospitable soil for a while, plants with root nodules leave enough nitrogen compounds in the soil to support a new wave of plant settlers that have neither root nodules nor the ability to supply their own nitrogen compounds.

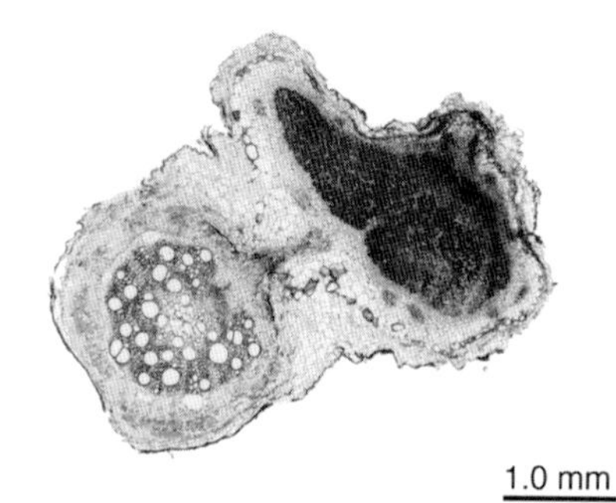

6. A cross-section of a root and its nodule. Bacteria live in the dark core of the nodule.

D. PLANT ROOTS AND THEIR FUNGAL PARTNERS

Although the roots of only a relatively few plant species and even fewer plant families team up with rhizobial bacteria and actinomycetes to produce root nodules and nitrogen compounds, the roots of all types of plants, from mosses to trees, have teamed up with long strands of fungi in special partnerships that benefit both plants and fungi. Together the two partners retrieve elements from deep in the soil and eventually return the elements to the topsoil where these nutrients once again begin their journey downward. Long strands of fungi surround plant roots and usually extend well beyond the reach of the longest plant roots. Since the thin fungal strands can also grow into tinier spaces of the soil than can plant roots, the fungi can tap resources that are otherwise inaccessible to roots, sending back water and a variety of minerals to the plant. And the plant reciprocates by sharing with fungi the energy of sunlight as well as the sugars that it makes with this energy. At certain seasons fungi use some of this energy from their green partner when they send forth their fruiting bodies, or mushrooms (plate 5). Thousands of fungal filaments from the soil come together to form these colorful mushrooms. The thousands of spores that each mushroom produces are carried off by wind and animals of the forest floor to new sites where they settle in the leaf litter, sprout their filaments, and start new partnerships.

It is easy to imagine how plants and fungi first became steadfast partners. Early, barren soils were most likely lacking organic matter as well as many elements, like nitrogen, phosphorus, and sulfur, that are found primarily in organic matter. In extracting elements from those soils, green plants probably needed all the help they could get, and fungi turned out to be committed allies. Fungi have been found in association with the first green plants that colonized the land about 400 million years ago. The development of partnerships between green plants and fungi might just have been a critical step in the early colonization of soils by green plants.

Fungal strands are particularly good at recovering phosphorus from the soil. Phosphate represents a compound made up of phosphorus and oxygen that is negatively charged and that neither leaches from the soil nor moves about much in the soil. Roots and fungi must locate whatever supplies of phosphates exist locally and not depend on slow-moving phosphates eventually reaching them. If it weren't for their fungal associates, green plants would not colonize acidic soils with little or no phosphates such as those of strip mines and peat bogs. In these phosphate-poor soils, roots associated with fungi can recover phosphates from a much larger volume of the soil than can plant roots without the help of the fungi. At best, any plants without fungal partners that manage to take root in these soils are small and stunted (fig. 7).

There was a time when soil fungi and moldy soil were considered harmful to plants. Since the few soil fungi that had been studied at the time did cause diseases of plants, people had the view that all fungi were harmful to plants and impeded their growth. How a few careful observations can change our views of the world! Not only do many soil fungi improve the health of plants that grow with them, but some of these fungi are actually essential for survival of the plants with which they associate, as a professor of biology in Berlin first proposed in an article published in 1885.

Professor A. B. Frank had originally been commissioned in 1881 by the German government to find a way to increase the supply of the delicious and highly prized fungi called truffles that grow among the roots of beech and oak trees. For centuries pigs and dogs have been trained to locate these woodland fungi, which fetch a high price at market, and German politicians naturally wished to encourage this lucrative harvest in their country. Although Frank never did find out how to improve Germany's truffle harvest, he found something very unexpected and even more important. Almost all the trees that he examined showed no evidence of damage even though molds and fungi encircled their roots. The fungi formed regular networks around the healthy roots. The intimate association of soil fungi with roots represents neither an ordinary root nor an ordinary fungus but a struc-

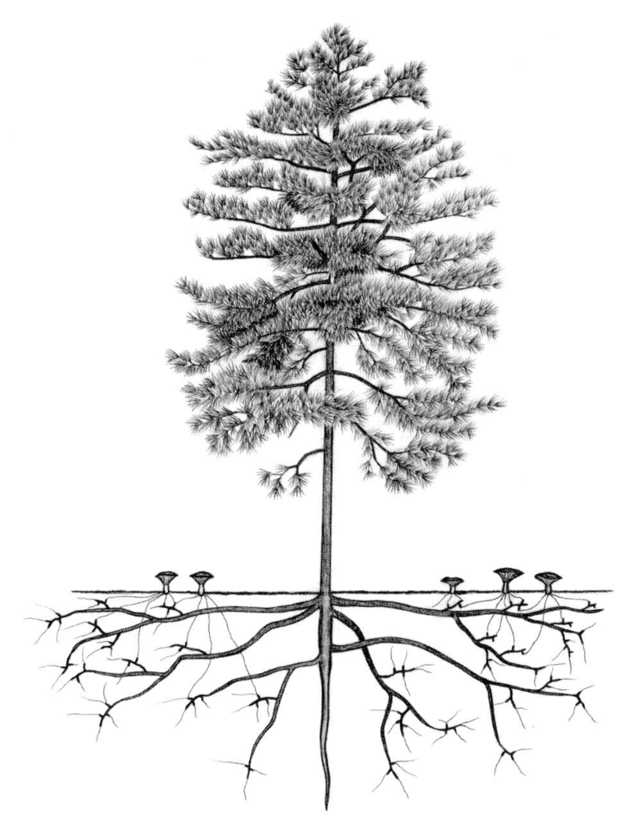

7. Fungal filaments wrap around roots of trees and exchange water as well as minerals for the sugars that the tree produces. The fungus uses the nutrients it obtains from the tree to produce its fruiting bodies, the mushrooms.

ture with properties of both. Since those early investigations, these special structures of the soil have been referred to as mycorrhizae or fungus (*myco*) roots (*rhizae*) and come in two forms: endomycorrhizae and ectomycorrhizae (*endo* = inner and *ecto* = outer).

The fungi of endomycorrhizae, or vesicular-arbuscular mycorrhizae, have hyphae that actually penetrate the walls of root cells but not the membranes of the cells (fig. 8, plate 6). As a hypha passes through a cell wall, the adjacent cell membrane remains intact and molds its shape to that of the invading fungus as the hypha assumes either a spherical, vesicular form or a highly branched arbuscular (*arbor* = tree; *-culus* = little) form that resembles a little tree. Nutrients are ex-

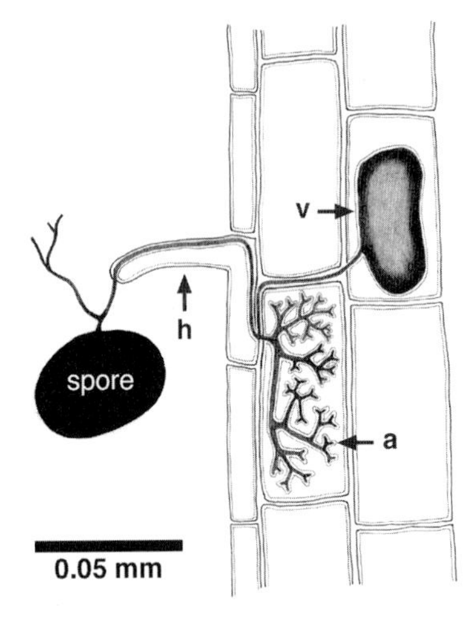

8. This diagram of an endomycorrhizal relationship shows how the fungus (shaded) and root cells interact and exchange nutrients without harming either one's cell membranes. For this vesicular-arbuscular mycorrhiza, a root hair (h), a vesicle (v), and an arbuscule (a) are shown.

changed at the arbuscular interface between the fungus and the membrane of the root cell, and nutrients are stored in the vesicles. Even though only about 130 species of endomycorrhizal fungi have been described, they are believed to form alliances with as many as 300,000 species of plants. These fungi are all in the phylum Zygomycota, order Glomales, and can grow only in association with plant roots.

About 5,000 species of fungi form ectomycorrhizal alliances with about 2,000 species of conifers and flowering trees. Practically all these fungi also form mushrooms and are members of the phylum Basidiomycota; only a few are members of the phylum Ascomycota. Unlike the fungi of endomycorrhizal alliances, however, practically all these fungi can grow independently of tree roots. Several layers of ectomycorrhizal hyphae surround each small tree root and radiate into the surrounding soil. These hyphae also penetrate roots by passing between cell walls, establishing a network of filaments among the cells of the roots without ever penetrating their cell walls or cell membranes (plate 7).

Professor Frank had enough confidence in his early observations to claim that mycorrhizae benefit plants by conveying water and nutrients to their roots. For years other scientists challenged his view that the association of fungi and roots was a healthy association and not a harmful one. Investigators have now traced the flow of nutrients between roots and fungi. They have also shown that pine seedlings with mycorrhizae absorb more nitrogen, more potassium, and more than twice the phosphorus from the soil than do pine seedlings without fungal partners planted in the same soil. Fungi and plants grow up together, sharing nutrients, but always respecting each other's integrity.

Ironically, although thousands of papers and books have been published about mycorrhizae since Professor Frank's pioneering work, no one yet knows how to increase the truffle population in an oak or beech forest.

E. WHERE ROOTS MEET ROCKS AND MINERALS

Once established in the soil with help from their pioneering partners—the algae, lichens, mosses, bacteria, and fungi—plants with roots can manage very well at living off the land and at the same time enriching the earth. Green plants contribute to the process of soil formation—by breaking down rocks and adding mineral matter to a soil as well as by adding their organic matter to the soil whenever they return in part or entirely to the soil. Their roots not only exert forces strong enough to crack rocks, but they also release compounds potent enough to gradually and inexorably transform some of the largest and hardest rocks to the finest of mineral particles (plate 8).

In its early days, a plant puts most of its energy into growing down. Enough nutrients are stored in its seed to tide the seedling over for a few days until its leaves start to form and expand. Roots, however, are the first part of a new seedling not only to form but also to function in carrying water and nutrients from the soil to parts aboveground. Small roots branch from large roots, and tiny root hairs branch from small roots. A typical Midwestern prairie produces three tons of roots for every ton of shoots. Through the months of the winter, roots are often the only parts of plants that live on. Even after plants are well established aboveground and belowground, the roots of most plants actually outweigh and outgrow all the leaves and stems, flowers and fruits that the plants produce aboveground (figs. 9–10).

As roots navigate the soil, they force their way into whatever passageways they encounter between particles of soil, often pushing soil to one side as they forge ahead. Look at a sidewalk that has been built over roots of trees to see just how forceful a growing root can be. An introductory botany text written in 1925 cites an example of the roots of a birch tree entering a crevice and lifting a twenty-ton boulder. By exerting a force of up to 150 pounds per square inch, a root can lift

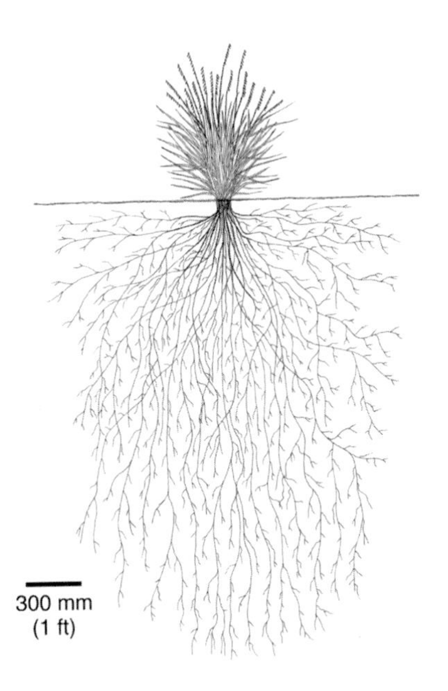

9. The roots of side oats grama grass spread deep into the prairie soil.

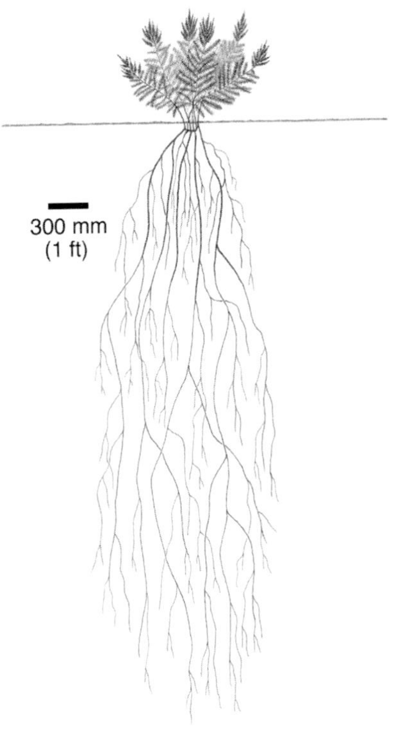

10. The deep roots of leadplant improve the structure of soil as they grow and enrich soil whenever they die. Each year prairie plants add to each acre of soil about three tons of roots for every ton of shoots.

and even crack a slab of concrete weighing hundreds of pounds. Or to put it another way, a root measuring four inches in diameter and three feet in length can lift 50 tons.

High above the ground and the roots, leaves capture the sun's energy to produce sugars and, like all other living tissues, release carbon dioxide gas. Roots—whether large, small, or tiny—may not directly capture the sun's energy, but they still need some of this energy for their survival. As roots break down sugars and release energy for their own use, they also release carbon dioxide into the surrounding soil just as leaves at the other end of the plant release carbon dioxide into the surrounding air. Each root is surrounded by a halo of carbonic acid that forms as carbon dioxide leaves the root and reacts with water in the nearby soil. The carbonic acid reacts with rocks like granite to release nutrients for plant growth and to form clay and silt. This simple acid has far-reaching effects on the future of the soil and the mineral nutrients that the soil can furnish to the roots of plants.

In a prairie, forest, or desert, the roots of different plants partition the soil space so that competition for nutrients and water is minimized. Some plants have shallow roots that rarely extend more than a few inches or a couple of feet below the surface of the soil. In deserts, where only about three percent of the rainfall penetrates more than a few inches belowground, cacti of all sizes—from giant saguaros to small, round cacti not much bigger than a nickel—have shallow, sprawling root systems that catch and absorb whatever rain happens to fall.

The roots of other plants extend down well beyond two feet, some as far as five feet; but even more roots, as in the case of desert mesquite trees, venture well beyond five feet belowground to depths of as much as 100 feet. Roots of some prairie plants, as well as trees like mesquite that live in some of the driest sections of a desert, extend belowground more than ten times as far as their leaves and stalks reach aboveground (plate 9).

As young seedlings, plants devote practically all their energy to growing underground. They expend relatively little energy growing toward the light until they have finally discovered a reliable source of water. Water and mineral elements beyond the reach of shallow roots may still be within reach of roots that extend deep into the soil. By having roots that extend to different layers of the soil, plants in a community reduce their competition for resources like water and minerals that may be scarce in certain soils.

Roots grow and travel through soils of many textures, wending their way among soil particles and along tunnels of earthworms, always in search of the water and oxygen that every root must have. Roots are capable of growing to phenomenal lengths. After only four months of growth, one rye plant had extended fifteen million roots, and the total length of these roots was about 380 miles. In terms of surface area, these roots came into contact with 2,554 square feet of soil. And if the lengths and surface areas of all the minuscule but innumerable root hairs covering each root are taken into consideration, the figures for root length and surface area soar to 7,000 miles and 7,000 square feet.

Having roots with such incredibly large surface areas assures that a plant gathers enough of the mineral nutrients that are present in low concentrations in the surrounding soil. By establishing mycorrhizal associations, roots increase even more the surface area available for uptake of water and nutrients. While roots are alive, they help hold soil particles together. Even after they die, roots leave a legacy to the soil that nurtured them. New spaces are continually being created in soil whenever roots rot and leave their vacant passageways in the soil.

Alive or dead, roots actually improve the movements of water, air, and other roots through the soil by helping to form clumps and aggregates of small particles of soil. However, because roots are so hard to observe deep underground, very few people have bothered to study them. When they do, scientists will probably discover that roots of many plants grow much longer and much deeper than they originally suspected.

Roots can be very choosy about the soils they associate with; they grow well only in soils that have a suitable assortment of nutrients. And often the nutrients can be used only if suitable mycorrhizal fungi are present in the soil to help the roots tap these resources. The 18 essential elements that plants need for growth and survival come from the air and from water, but mostly from the soil. The three elements derived mostly from water and air—carbon, hydrogen, and oxygen—make up 95 percent of a plant's weight. The other 15 essential elements derived mostly from the soil represent a mere 5 percent of a plant's weight.

Elements and compounds in the soil can occur in uncharged forms, but most often occur as either positively charged ions (cations) or negatively charged ions (anions). Like charges repel and opposite charges attract. When negative charges combine with an equivalent number of positive charges, the resulting compound is uncharged or neutral as happens, for example, when a calcium cation meets a carbonate anion.

Ca^{+2}	+	CO_3^{-2}	→	$CaCO_3$
calcium cation		carbonate anion		calcium carbonate or lime

A number of elements—notably nitrogen, phosphorus, sulfur, and boron—can be used by plants only when they form compounds with hydrogen such as ammonium cations (NH_4^+) or with oxygen such as nitrate (NO_3^-), phosphate (PO_4^{-3}), sulfate (SO_4^{-2}), or borate (BO_4^{-2}).

Even though positive cations and negative anions attract and bind one another, under certain circumstances they can separate, exchange, and form new combinations. Exactly which exchange of cations and anions will occur in the soil will depend, first of all, on how strongly the anions and cations bind to each other and, second, on how many there are of each. Compounds and elements interact in the soil to either enhance or suppress each other's uptake by plant roots or movement in the soil. If one compound or element in a soil increases or decreases, other compounds or elements, as well as the plants and animals that live in that soil, are often affected.

The chart below lists the elements that are essential for plant growth along with the ionic forms that they assume in soils.

The first nine elements are those that are required in relatively large amounts (more than 0.1 percent of a plant's dry weight); the remaining elements in the list are those that are used in relatively small quantities (less than 0.1 percent of a plant's dry weight).

Names and symbols for elements	**Names and symbols for ionic forms of elements**
Elements derived mostly from water and air	
Carbon, C	
Hydrogen, H	H^+
Oxygen, O	

Elements derived mostly from soil

Nitrogen, N	NO_3^-, NH_4^+	(nitrate, ammonium)
Phosphorus, P	HPO_4^{-2}, $H_2PO_4^-$	(orthophosphates)
Potassium, K	K^+	
Calcium, Ca	Ca^{+2}	
Magnesium, Mg	Mg^{+2}	
Sulfur, S	SO_4^{-2}	(sulfate)
Boron, B	BO_4^{-2}	(borate)
Copper, Cu	Cu^{+2}	
Chlorine, Cl	Cl^-	(chloride)
Iron, Fe	Fe^{+2}, Fe^{+3}	(ferrous, ferric)
Manganese, Mn	Mn^+	(manganous)
Molybdenum, Mo	MoO_4^{-2}	(molybdate)
Zinc, Zn	Zn^{+2}	
Nickel, Ni	Ni^{+2}	
Cobalt, Co	Co^{+2}	

Roots mine the mineral resources of the soil, and roots make these 18 essential nutrients available to animals of the land. In going from soil to plant, calcium, for example, is first concentrated eightfold; and in going from plant to animal, this element is then concentrated fivefold more, for a total of 40 times the concentration found in soil. Animals are not only entirely dependent on energy that is ultimately derived from sugars produced by plants during photosynthesis, but also almost entirely dependent on plant roots for concentrating from the soil the many mineral nutrients that sustain their lives.

To establish how much nutrition a plant gets from the soil, the Belgian scientist Jan Baptista van Helmont performed a simple experiment over 350 years ago that we still cite today as a particularly informative experiment on plant nutrition. First, van Helmont filled a large pot with exactly 200 pounds of soil that he had thoroughly dried in an oven. Next he watered the soil and planted a shoot that he had cut from a willow tree. He weighed the willow shoot at the time he planted it and then again five years later.

The shoot that initially weighed five pounds grew into a small tree weighing 169 pounds and three ounces. Throughout the experiment, he only added water to the soil to keep it moist, and he carefully covered the surface of the pot with a sheet of metal perforated with tiny holes to keep dust from accumulating on the surface of the soil. In addition to weighing the tree after five years of growth, van Helmont also dried the soil from the pot and found that the soil had lost only two ounces of its initial starting weight of 200 pounds. Most of the 164 pounds that the willow shoot had gained in the five years must have come from water and air, but only a tiny fraction of the plant's increase in weight could have come from the soil.

One of the most illuminating demonstrations about the nutrition that plants derive from soil is described in an engaging little book written in 1950 by Sir John Russell, who was then director of the Rothamsted Experimental Station in England. *Lessons on Soil* describes a simple experiment that Russell designed for students at the village school of Wye in Kent, England. Expanding on what van Helmont had shown earlier about how little food a plant actually receives from soil during its life, Russell set out to show that the remains of once-living plants represent an important contribution to these small, practically undetectable amounts of nutrients from the soil—nutrients without which plants barely survive and grow only poorly.

Russell asked the students to plant mustard seeds in a pot of topsoil where rye plants had previously grown (pot 1). The mustard plants that subsequently grew weighed only 17.8 grams while mustard plants that grew in topsoil where no other plants had grown (pot 2) weighed 62.3 grams—three and a half times as much as the mustard plants that had shared their soil with rye plants. Van Helmont would have predicted exactly what the students found when they dried and weighed the soil of pot 1 before planting the rye and after harvesting it. Even though the rye plants had used most of the nutrients in the soil of pot 1, the dried soil neither gained nor lost weight during the time that the rye plants grew on it (plate 10).

When the students repeated the experiment using subsoil rather than topsoil, mustard plants that had grown in subsoil where rye had previously grown (pot 3) weighed slightly less than mustard plants grown in soil that had not been previously planted with rye (pot 4). Although topsoil provides nutrients for growth of mustard plants, subsoil offers less in the way of nutrients to either rye plants or mustard plants.

To help pinpoint the origin of these nutrients in the soil, the students were then asked to compare the growth of mustard plants in specially prepared pots of topsoil (pot 5) and subsoil (pot 6). To both of these pots, the same weight of such plant remains as stems, leaves, or grass were added before the mustard seeds were planted. The mustard plants always grew larger in both topsoil and subsoil to which plant pieces had been added (pots 5 and 6) than they did in soil to which no plant remains had been added (pots 2 and 4). Once again the mustard seedlings grew better in the pots with topsoil (pots 2 and 5) than in the pots with subsoil (pots 4 and 6). Adding pieces of plants to the subsoil (pot 6) increased the growth of mustard plants, but not as much as adding the same weight of plant pieces to topsoil increased growth of the mustard plants (pot 5).

The plant pieces that the students had added to the soil provided nutrients to the growing mustard seedlings, but the plants were not able to take nutrients directly from the pieces of plants. If the mustard seedlings had been able to obtain nutrients this way, they would have grown as well in subsoil containing plant remains (pot 6) as they grew in topsoil containing the same amount of plant pieces (pot 5). When the pots were emptied after the mustard plants had completed their growth, more original plant pieces remained in the subsoil than in the topsoil. Some change occurs in pieces of plants when they are exposed to the soil that makes them more nutritious to growing plants. Whatever this change is, it is more evident in topsoil than it is in subsoil; and the agents that bring about this change are clearly more effective in the topsoil than they are in the subsoil.

These experiments that the students carried out call attention to the special attributes of topsoil. One obvious difference between topsoil and subsoil is that topsoil is inhabited by creatures whose chief occupation is decomposing and recycling plant and animal remains that fall to the surface of the soil. These creatures are busiest and most abundant in the topsoil, because it is here that their food and nutrients are most plentiful. The fertility of a soil comes from the life that it supports.

F. PLANT ROOTS AND THEIR ANIMAL PARTNERS

1. Life in a Dark and Densely Populated World

Many of the animals of the soil look like creatures from another world, and they really are creatures from a world that is totally alien to most of us. Few people ever see these creatures, not because they are rare but because they lead retiring lives in the leaf litter and soil.

Try looking for them, however, and you are very likely to succeed in your search. Despite their uniqueness, these small animals without backbones (*invertebrates*) share a number of features with other animals of the soil, both large and small.

A great many of the soil creatures are pale and white. In the dark recesses of the soil, pigments that impart color to animals are embellishments that none of their blind fellow creatures would appreciate anyway. Pigments often protect from sunlight; but in the soil, pigments are usually superfluous since rays of sunlight almost never reach these recesses. There is no point in wasting energy on producing pigments that serve no useful function.

In totally dark surroundings there is also not much point in having eyes. Since eyes are often absent or very small in soil creatures, senses other than sight have been enhanced to compensate for visual deficiencies. Touch and smell seem to be particularly acute. One feature of most creatures from the underground is their abundance of touch-sensitive hairs, bristles, and whiskers. These sensory structures can be quite large and can come in any number of shapes: some long, fine, and tapered; some like fine feathers; some paddle-shaped; and a few club-shaped.

Predators are usually endowed with well-developed sense organs since their livelihoods depend on finding particular prey. A good sense of touch and smell go well with speed and agility in tracking down prey. In the labyrinthine passageways of the soil, long, sinuous bodies are often a definite advantage for a predator. Nonpredators often just move along until they stumble upon their dinner—sometimes living, sometimes dead.

Rather than trying to outrun or outmaneuver the agile predators, prey often use their energy to form hard, protective shells that are part of their bodies. If these shells are not sufficient protection, the prey often curl or roll into a ball with all their soft, vulnerable parts tucked away beneath the ball's hard surfaces.

Mites, millipedes, woodlice, and armadillos have all independently discovered this strategy for protection. When curled into a ball, one genus of woodlouse looks so similar to a curled armadillo, the Spanish word meaning "little armored one," that the person who initially named the woodlouse decided to name it *Armadillidium* (figs. 11–12). Oribatid mites of soils also have hard, slick shells covering most of their spherical bodies. A hungry predator finds it difficult to sink its

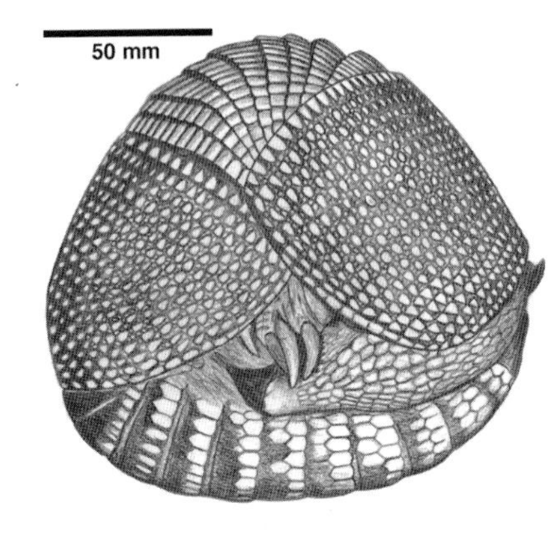

11. The armadillo belongs to a group of armored mammals that dig burrows and feed on soil insects. These slow-moving creatures escape the teeth of their predators by rolling up so that only their hard armor is exposed.

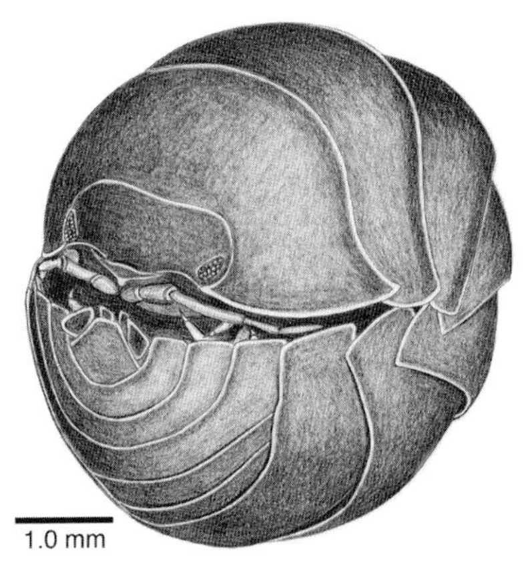

12. The *Armadillidium* is a slow-moving woodlouse that avoids the jaws of its predators by tucking all its appendages beneath its hard plates of cuticle.

mandibles into the protective shell, but the mite's eight legs are still vulnerable bite-size morsels. These legs can, however, be drawn beneath the protective shell at the rear end of the mite while the hard shell at the head end that is hinged to the rear end swings backward. The two shells clamp down like those of a clam, with barely any space between them (see fig. 53). By thus rolling into a ball, the oribatid leaves no vulnerable parts of its body exposed. Predators are foiled by this maneuver, and the mites are also protected from desiccating if the soil ever becomes too dry.

Other animals of the soil protect themselves by building cases in which to retreat or lay their eggs. A few beetle larvae and a few moth caterpillars use a combination of silk, saliva, and soil particles to fashion the cozy cases in which they live and travel. The rear end of a case is usually open to expel the droppings of the larva, and the larva can

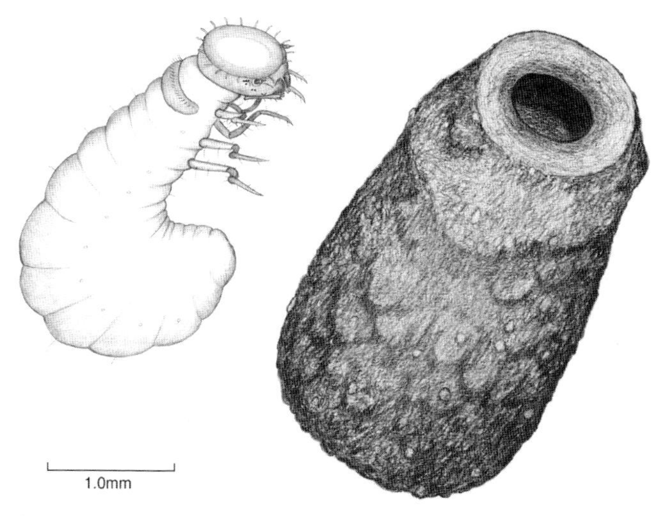

13. The head of this beetle larva forms a snug-fitting plug for the case that it fashions from its droppings and shed skins.

quickly seal the front end by plugging the entrance to the case with its very hard, snug-fitting head (fig. 13).

Although millipedes and pseudoscorpions do not carry cases wherever they go, they build shelters for themselves at each molt, a time in their lives when they are most vulnerable. As they grow, their old, hard shells are shed, exposing new, soft shells underneath. Several days may pass before the new shell expands and hardens. A mother millipede also builds a case for each of her eggs and then stands guard over her nest as the young millipedes develop inside. Cases made from particles of soil defy the jaws of predators and blend in so well with surrounding soil that they easily go unnoticed.

Another feature shared by many of these soil animals is their sheer abundance. Among the most abundant soil animals are the arthropods (*arthro* = jointed; *pod* = feet)—animals that have jointed legs and no backbones. Well over a billion arthropods may live on an acre of pasture soil. In a hardwood forest, the layers of leaf litter and the tiny measureless passageways that permeate the rich soil provide even more habitats for arthropods of the soil.

The figures that have been calculated for the number of individuals of a particular group of soil creatures may vary depending on the time of year the population was examined, the particular habitat occupied by the population, and the method used to count portions of the total population. Since figures calculated for human populations can also vary with season and place, taking a census of a country with millions of people is a monumental undertaking. The final count of people represents not an exact count but nevertheless a good estimate of the actual size of the population.

Conducting a census of organisms on an entire acre of soil can also be a demanding effort. Census takers counting at different times and places have arrived at different figures for the numbers of organisms per acre of soil. The figures, however, always give the same impression: the organisms of the soil are extremely abundant, and certain groups are always more abundant than others. The number of creatures on a single acre of land far exceeds the entire human population of the world.

Think of organizing all the animals, bacteria, and protozoa in a square meter or square yard of forest soil or prairie soil into categories according to their abundance. We can do this by arranging the categories as a series of stacked layers with the dimensions of each layer reflecting the number of creatures in a particular category. When the layers are stacked in order of their sizes, they form a pyramid of numbers (fig. 14). The most abundant organisms will be represented by the very bottom layer, and the least abundant will be represented by the layer at the very top of the pyramid.

Even though they are the smallest of the soil creatures, bacteria can easily claim to be the most abundant—on the order of thousands of billions in each square meter of soil. There are so many bacteria that despite their individual minuscule sizes, as a group they easily outnumber, and often outweigh, all other categories of soil creatures.

As soil creatures increase in size (fig. 15), fewer of them are found in a square meter of soil. So by ranking the categories of creatures according to their sizes, we can predict the arrangement of layers in the pyramid. The one-celled organisms known as protozoa would be represented by a layer above that for the smaller, more numerous bacteria but below the layer for the larger, less numerous nematodes. Then seven more layers—the first six of which represent (1) mites, (2) relatives of insects known as springtails, (3) tiny invertebrates known as rotifers and tardigrades, (4) all arthropods other than mites and springtails and including insects, (5) potworms and earthworms, and

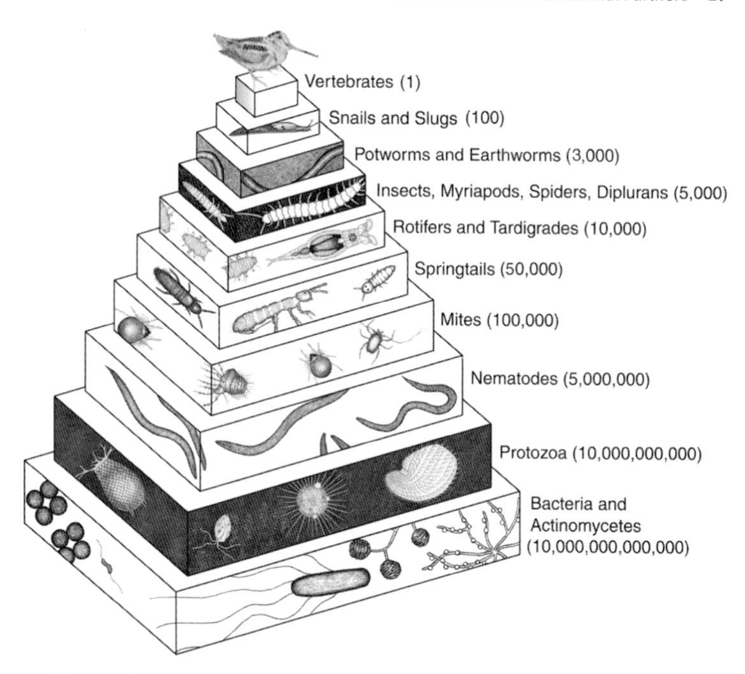

14. The number of animals, protozoa, and bacteria that live on a square meter of ground can be arranged in a pyramid of layers according to their sizes. There are millions of creatures at the bottom layer of the pyramid of numbers, a few in the top layer, and thousands of invertebrates in layers in between. If the dimensions of each layer in the pyramid were drawn proportionally to the number of creatures that they represent, then the layer for bacteria and actinomycetes would be 100 million times the volume of the mite layer; and the mite layer, in turn, would be 100,000 times the volume of the vertebrate layer.

finally (6) snails and slugs—would be stacked one on top of the other in the order of their increasing sizes.

The topmost and smallest layer of the pyramid would represent the vertebrates—toads, salamanders, turtles, snakes, mammals, and birds. These larger animals may dispatch large numbers of earthworms, insect larvae, and slugs; but the pyramid on which they are the topmost layer neither wobbles nor tips. The vast mosaic of soil life needs predators as well decomposers, minuscule creatures as well as hefty, all subsisting in an exquisite state of balance and sustaining one another. The interactions of all these members of the soil community take the form of an intricate web, showing how the various members depend on one another for their nutrition and survival (fig. 16).

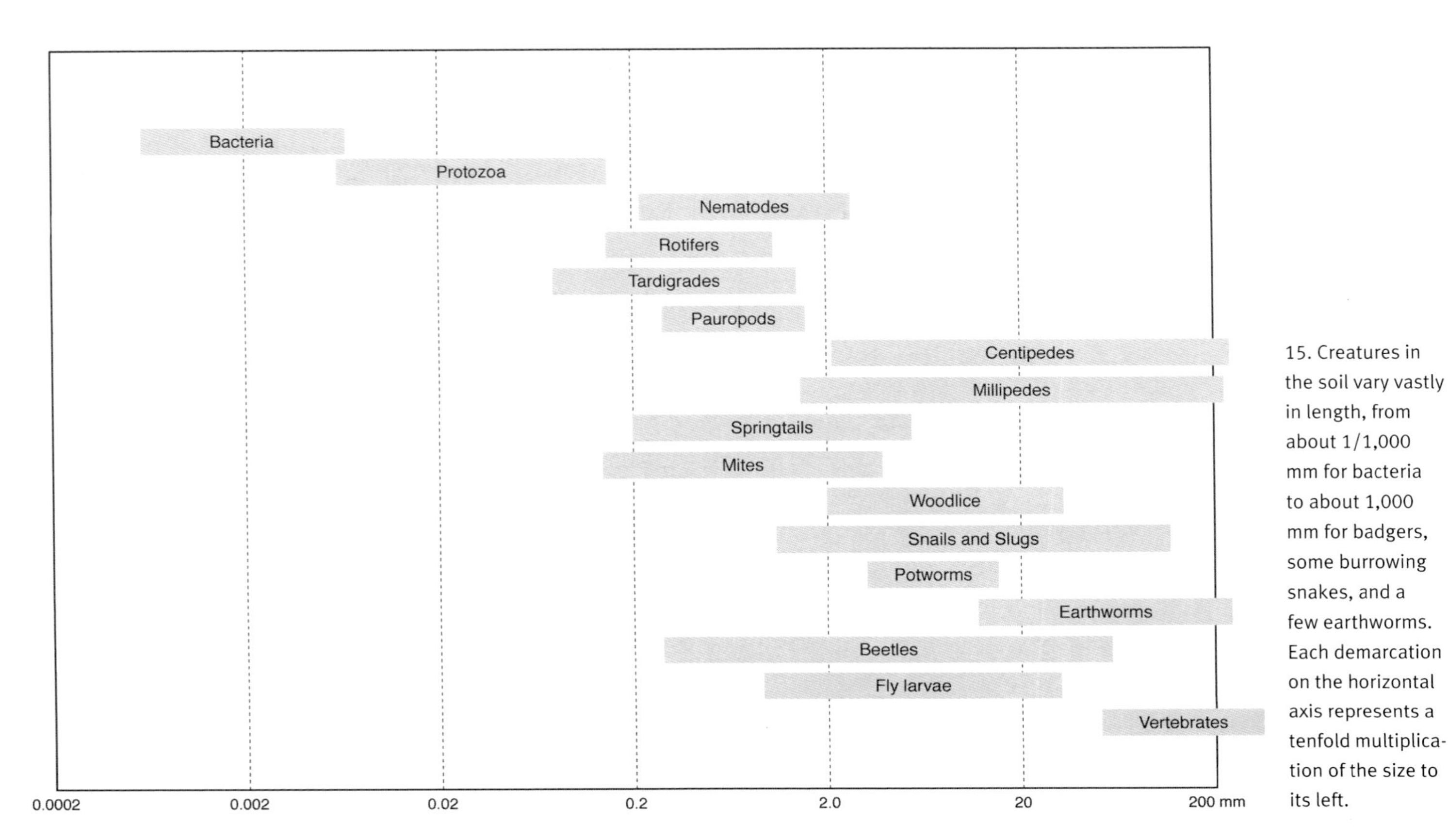

15. Creatures in the soil vary vastly in length, from about 1/1,000 mm for bacteria to about 1,000 mm for badgers, some burrowing snakes, and a few earthworms. Each demarcation on the horizontal axis represents a tenfold multiplication of the size to its left.

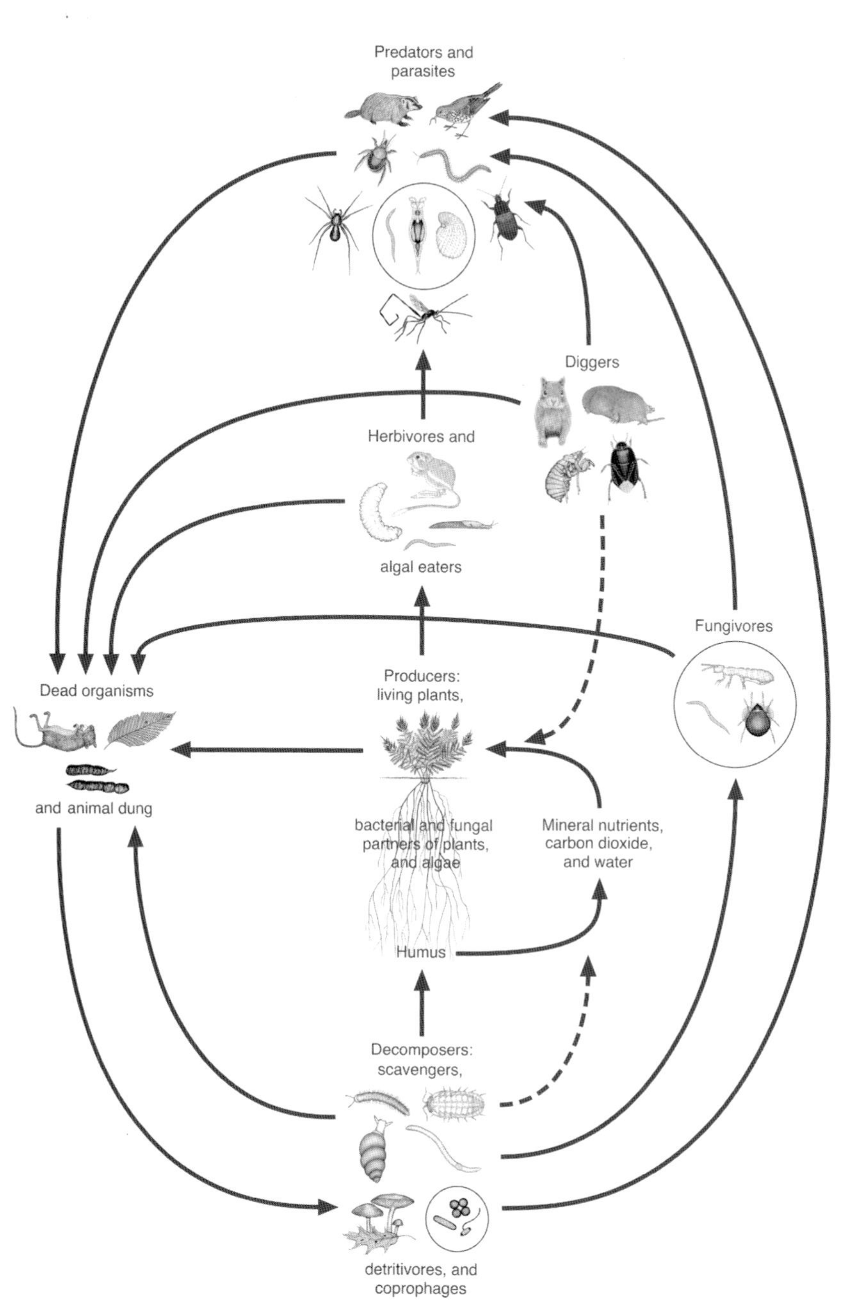

16. This food web shows how both living and dead creatures interact with each other as well as with minerals, water, and carbon dioxide of the soil. Solid arrows show the flow of nutrients between the living world and the mineral world of the soil; and dashed arrows indicate how the flow of nutrients from debris and soil is influenced by the activities of decomposers, scavengers, detritivores, and diggers.

2. Soil Fertility and the Formation of Humus

As he watched leaves fall gently to the earth one autumn day, Henry David Thoreau clearly realized that even death and decay have a grander and brighter side than most of us appreciate (plate 11).

> How pleasant to walk over beds of these fresh, crisp, and rustling fallen leaves . . . How beautiful they go to their graves! They that waved so loftily, how contentedly they return to dust again and are laid low, resigned to lie and decay at the foot of the tree and afford nourishment to new generations of their kind . . . They are about to add a leaf's breadth to the depth of the soil. We are all richer for their decay.

The people of the Aran Islands, off the west coast of Ireland, have known for centuries that to grow vegetables on their rocky islands they must pulverize their rocks and mix them with decaying seaweed. Not much but algae and lichens and mosses can grow on rocks and minerals alone. The organic matter contributed by plants and animals enriches a soil's physical as well as chemical properties, providing the essential conditions for survival of practically all of the soil's plant and animal life.

All living plants and animals take elements and compounds from the soil. In death they actually return more to the soil than they ever took during their lifetimes. Not only do they return elements and compounds that they obtained from the soil, but they also add organic matter made up almost entirely of elements that initially came from water and air. Much of the energy that plants capture from the sun during photosynthesis is passed on to the soil and to the animals that feed on the plants.

Plants are nourished by resources of the soil, and they eventually return all elements to the soil, but not without a great deal of help from the decomposers. In addition to returning to the soil all the elements that plants once borrowed from it, decomposers inherit the energy that plants once captured from the sun as they grew and flourished. The decomposers pass on their inheritance to the soil as they actively decompose the remains of plants and as they eventually are in turn decomposed by other decomposers. This energy from a faraway sun is put to use by the decomposers of the soil as they break down dead plant and animal matter.

Traces of plant and animal matter often stubbornly resist decay. The tough and fibrous traces of plant remains like leaf veins and heartwood are made up of sugars strung together to form chains of cellulose and hemicellulose as well as compounds known as lignins, which hold the chains of sugars together. A few creatures of the soil, like snails, termites, and fungi, can digest some of these tough plant remains, but most decomposers simply pass them through their guts undigested, but at least chewed down in size.

Even though they decompose very slowly, these tough plant remains become smaller and smaller as they pass through one soil animal after another. Organisms in the soil eventually pulverize these persistent plant fragments even though the cellulose and lignin making up these fragments remain mostly intact. By the time bacteria have traveled through the guts of larger decomposers, they have often multiplied many times over, and they continue breaking down whatever organic matter remains in the droppings of the decomposers.

After organic matter passes through an earthworm, for example, its bacterial population can increase fivefold. Droppings made up of these pulverized fragments of leaf litter are even more inviting to bacteria and fungi since there are now many more pores and crannies for them to occupy. The fungi that dwell in the forest litter are continually consumed in vast numbers and their spores are continually being carried to new homes by decomposers.

As microbes and animals of the soil break down the remains of various creatures into simpler compounds, they add organic matter to the soil. Such organic matter also happens to be the main source of certain essential elements of the soil such as nitrogen, phosphorus, and sulfur. Eventually the organic matter transforms to tiny particles of a dark organic matter known as *humus*. Humus is the hard-to-digest plant materials that remain in organic matter after the easily digested portions have been consumed at least once by decomposers. It is made up of those persistent organic molecules from plant cells such as oils, resins, lignins, and waxes as well as remains of bacteria and fungi that pass through the digestive tracts of decomposers and end up in their droppings.

Organic matter can be consumed a number of times, since there are some decomposers that actually prefer feeding on organic matter that has already been digested by other decomposers. Other soil arthropods and worms also find nutrition in these droppings. One creature's meal today can be another creature's meal tomorrow. These decomposers of dung, or *coprophages*, redigest the droppings of other soil animals. Among the better-known decomposers of dung are certain potworms, springtails, and woodlice. With time even the well-digested humus formed by the coprophages slowly and gradually decays. But while it lingers in the soil, it has some remarkable effects on soil properties.

Because humus particles are acidic and contain high concentrations of positively charged hydrogen ions, for each hydrogen ion lost, the humus particle gains a negative charge. The negative charges that accumulate on humus particles act as magnets for any positively charged elements or compounds (cations) that pass their way. Humus binds many of these cations that are essential for plant growth, like calcium and potassium, and prevents their being leached from the soil. Humus serves as an important intermediary between fresh organic matter, from which it is derived, and the carbon dioxide, water, and minerals to which it eventually returns, often years later.

The rate at which humus and organic matter of soils are converted to carbon dioxide, however, is linked to weather and soil conditions. Where soils are sandy and rocky, coarse and porous, as they are in deserts, organic matter decays more rapidly than it does in soil that is dense, compact, and poorly aerated. Organic matter also decays more rapidly in warmer climates. When temperatures rise above 80°F (25°C), organic matter actually decays faster than it is generated. That is why practically all organic matter in the tropics is found aboveground in the living forest. The bacteria and fungi quickly go to work on any organic matter that falls to the ground and reduce it to simple compounds of water, carbon dioxide, and minerals, leaving little debris or humus for larger creatures of the soil such as earthworms and insects.

The carbon dioxide that is released from organic matter and the humus of soil is a greenhouse gas that contributes to the troubling phenomenon of global warming. Only in the cooler forests at higher latitudes does humus have a chance to build up and slow the production of carbon dioxide. If global warming continues to increase, not only will the humus content and the richness of soils imparted by humus diminish, but also the levels of carbon dioxide in soil and air will undoubtedly rise.

3. The Importance of Nitrogen

Bacteria and fungi are the microbes that initiate recycling of dead plant and animal matter in the soil. Microbes have been around longer than other creatures of the soil; and as we are continually discovering, they have adapted to just about every environment and every diet. They can ingest, digest, and process just about any compound known to humans. That virtue has made microbes extremely useful in cleaning up many of our human-made messes as well as in processing substances in the soil that no other creatures can process. Without this processing, nature's chemical cycles would screech to a halt. Microbes get involved in the rather unsavory task of feeding on dead plants and animals in order to tap the reserves of energy that remain in plants and animals after death. Energy is a resource of unparalleled importance. All living things need energy for survival. Whenever the supply of dead matter that falls to the ground increases, microbes of the soil quickly multiply to use every bit of this new energy source.

Just as all creatures need energy for survival, they also need large amounts of certain elements—in particular, carbon and nitrogen, which are major constituents of their proteins and nucleic acids—to grow and multiply. Microbes can usually get all the carbon and other elements they need from dead plant matter, but nitrogen is often in short supply. The proportion of carbon to nitrogen for many fallen leaves and logs is greater than 100 to 1, but in bacteria this proportion is around 5 to 1. For these bacteria to survive and multiply, they need to find a source of food that can provide carbon and nitrogen in proportions ranging between 5 to 1 and 30 to 1.

As microbes undertake the job of decomposing fallen leaves or logs, they obviously need more nitrogen than the leaves or the logs can supply. For microbes to continue multiplying and decomposing, they must get nitrogen from whatever source is nearby. Microbes that feed on organic matter in its early stages of decay multiply and act most efficiently as decomposers if they can take nitrogen from the neighboring soil. This usually means competing with plant roots for nitrogen. The growth of plants in the neighborhood of decaying organic matter slows down as microbes compete with roots for the nitrogen that they need. The addition of nitrogen-rich manure or fertilizer to soil with decaying organic matter gives both roots and microbes enough nitrogen to satisfy their needs. Likewise, adding nitrogen to compost in the form of manure or fertilizer provides the missing element for microbial growth and speeds up the decay of compost heaps.

Fresh, green plant matter is higher in nitrogen than dry, dead plant matter like straw, sawdust, or dry leaves and is always a good addition to a compost pile. Once the decay of organic matter in soil or in compost is complete, and the energy source for bacteria is used up, most of the bacteria will die and return the nitrogen they used to grow and multiply so rapidly. Soil and compost are now the richer for the demise of the decomposers, and enough nitrogen is now available to meet the needs of plants and microbes alike.

Different plants and different parts of plants have different proportions of carbon to nitrogen, so naturally they differ in their ability to supply the nitrogen that microbes need to continue growing. Not surprisingly, those plants or plant parts that can offer decomposing microbes ample quantities of nitrogen rot faster than those with meager quantities of nitrogen.

Some leaves, like those of elms and alders, are thin and delicate; these leaves contain more nitrogen than leaves like those of oaks and beeches, which are thick and tough. Elm and alder leaves, in which the ratio of carbon to nitrogen is about 20 to 1, rot faster than oak, pine, and beech leaves, which have a carbon-to-nitrogen ratio of about 50 to 1; they decay in one year as opposed to three years for the tougher leaves.

Leaves like those of maple and basswood trees, with carbon-to-nitrogen ratios of about 40 to 1, are not quite as tender as elm leaves, or as tough as oak and beech leaves, and completely decay in two years. Sawdust, wood chips, logs, and branches contain an even smaller proportion of nitrogen than beech and oak leaves. Woody materials that contain far more carbon-containing compounds than nitrogen-containing compounds, with a carbon-to-nitrogen ratio as high as 400 to 1, rot the slowest of all the organic matter.

An oak leaf experiences many changes from the time it falls from a tree until the time it is transformed to particles of humus by the many decomposers of the forest floor. There are always plenty of springtails on the forest floor waiting to chew holes in this freshly fallen leaf, and they are often joined in this endeavor by woodlice and millipedes that also live in the leaf litter. Fungal filaments and bacteria invade the leaf tissue through the holes and improve the flavor of the decaying leaf for larvae of crane flies and midges. These fly larvae begin widening the holes chewed by springtails, woodlice, and millipedes, and add some of their own holes to the fallen leaf (fig. 17).

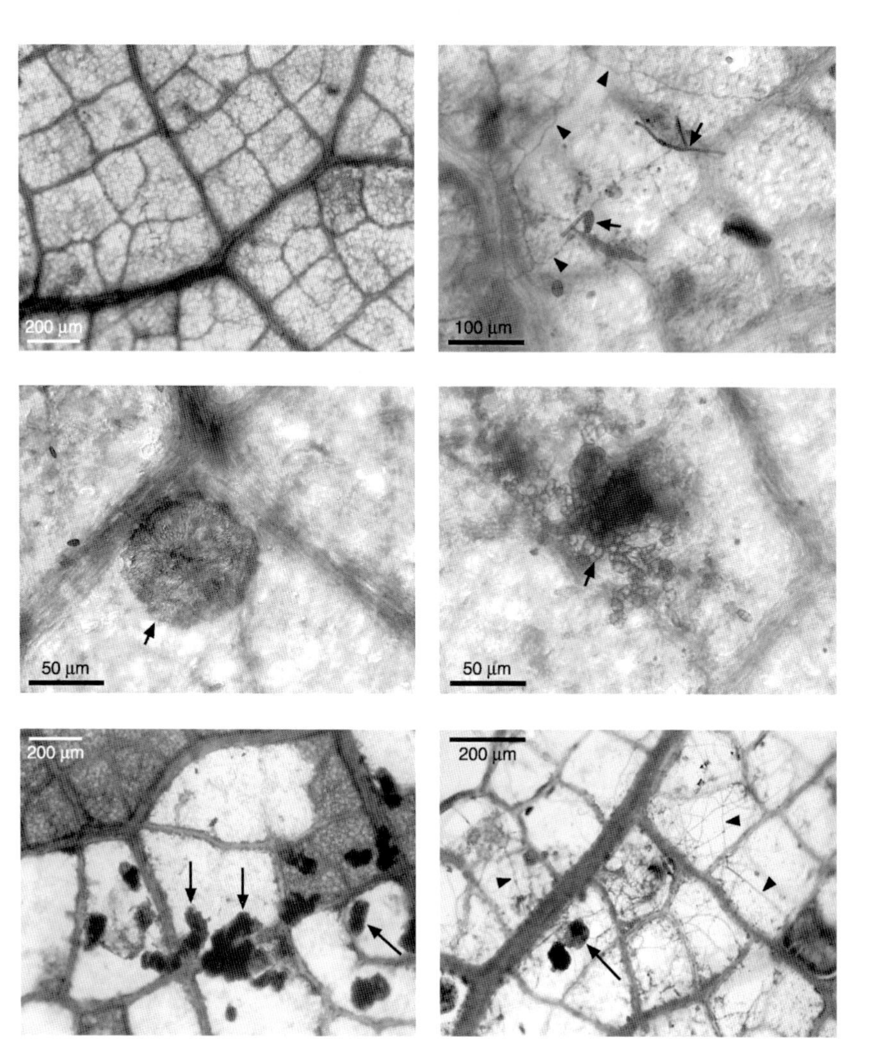

17. As oak leaves return to the earth, they pass through various stages of decay. They are colonized by fungi and microbes as well as skeletonized, shredded, and digested by arthropods. Short arrows indicate conidia or spores of fungi; arrowheads point to hyphae. The droppings that the arthropods leave behind (long arrows) contribute to formation of humus.

18. The skeletons of these hickory leaves are rich in lignin, a principal ingredient of woody fibers that is always the last portion of leaves and logs to decay.

At this stage, when the leaf is pulverized and perforated, even more soil arthropods join in the feast on the fallen leaf. Snails, oribatid mites, earwigs, a few crickets, other fly larvae, and maybe a few bristletails now widen the holes in the leaf until only its skeleton remains. Most of the soft, nitrogen-rich tissue of the leaf has been eaten, and only the tough, nitrogen-poor leaf skeleton or veins still persist. The many jaws that have chewed on the leaf have left many little pockets and irregularities in the leaf (fig. 18).

Bacteria move into these many microhabitats and now colonize more of the leaf surface than was ever exposed or colonized before. Bacterial decomposition progresses rapidly. Minuscule relatives of earthworms called enchytraeid worms or potworms, along with the ubiquitous oribatid mites and springtails, continue chewing what remains of the leaf's skeleton. Earthworms soon pull the remains of the fallen leaf into their extensive burrows, where they mix this organic matter with mineral matter from below.

Earthworms can be particularly abundant and are especially good processors of organic matter, but other burrowing animals also contribute to the mixing and churning of particles from the decomposed leaf. By this stage the leaf has been so thoroughly chewed and fragmented that only indigestible, dark, and enriching particles of humus remain. The decomposers have thus transformed a relatively nitrogen-poor oak leaf to relatively nitrogen-rich humus, and the soil is all the richer for the transformation.

4. The Contribution of Animals to Soil Structure

During a good portion of the nineteenth century, most chemists were convinced that understanding the soil's contribution to plant nutrition

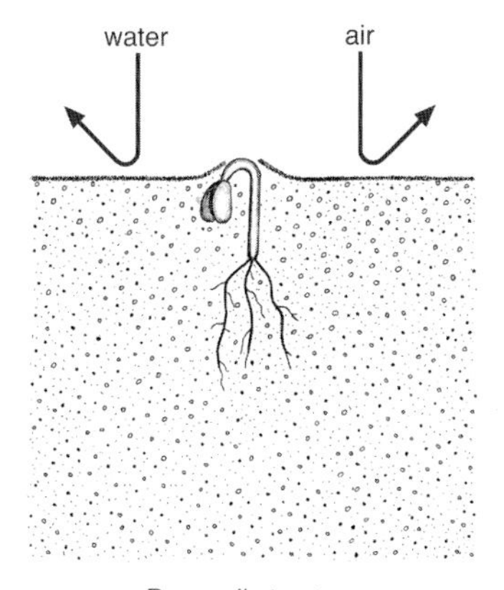

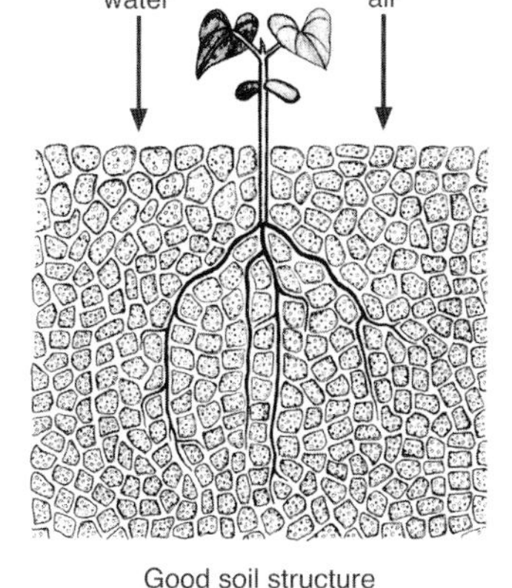

19. Seedlings emerge best from soil in which aggregates are organized in a hierarchy of sizes. These different sizes impart a highly desirable crumbly and granular structure to soil. Particles of sand (o), silt (•), clay and humus (·) are held together as aggregates by organic matter, algae, hyphae of fungi, and fine roots of plants. When aggregates fail to form or break down under heavy rain, compaction, loss of organic matter, or excessive fertilization and tillage, soil particles disperse and clog the pores that allow air and water to penetrate the soil. In this schematic diagram, soil particles, aggregates, pores, and plants are not drawn to scale. (Based on a figure from *Sustainable Soil Management* by Preston Sullivan, ATTRA publication #IP027/133)

was simply a matter of understanding which chemicals, such as nitrates or phosphates, should be provided to a plant and in what proportions. In the twentieth century it became increasingly clear that the growth of plants is influenced not only by the chemicals that make up the solid ingredients but also by the *structure* of the soil—the way in which its solid ingredients aggregate in the soil (fig. 19). By adding and mixing organic matter and humus with mineral particles, decomposers and recyclers encourage the formation of soil aggregates as well as improve the structure and, consequently, the fertility of soil.

Humus and the organic matter from which it comes can affect the structure of soil in at least two ways. They can hold soil particles together in aggregates and thus reduce erosion of what otherwise would be soil that is too loose. On the other hand, the presence of humus and organic matter can loosen the structure of a soil that is too tight and compact by facilitating movement of air and water into the

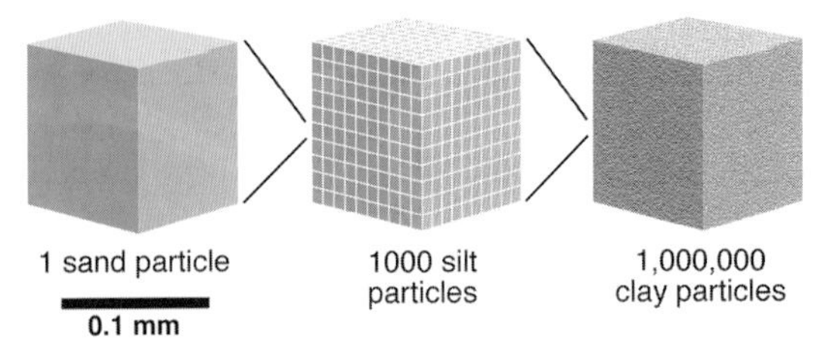

20. The surface areas for a given volume or mass of soil particles are related to the sizes of the particles that line the many pores (shown in white) of a soil. The widths of these sand, silt, and clay particles respectively measure around 0.1, 0.01, and 0.001 mm. The widths of silt and clay particles are around 10 and 100 times smaller than the width of sand particles, For a given volume of soil particles, the surface areas and the pores separating these particles also increase by at least a factor of 10 when comparing sand and silt. Pores and surface areas increase more than 100 times when comparing sand particles with clay particles.

soil as well as by encouraging the growth of roots and the movement of animals through the soil. This structure of a soil is often described as the size and shape of the small pieces into which a soil breaks up when crumbled in the hand. A granular, crumbly structure has spaces and pores among the solid ingredients of a soil through which air, water, roots, and soil creatures can freely move.

Although the solid ingredients appear to fill practically all the space occupied by soil with a crumbly structure, they actually share about half of their space with air and water. Tiny mineral particles of the soil like clay are separated by tiny pore spaces. Large mineral particles like sand are separated by large pore spaces. Even though larger sand particles may have larger pores between them, smaller clay particles have more of their surface area exposed to surrounding pores. Clearly, a mineral particle will increase its total area of exposed surface if it breaks into smaller particles (fig. 20).

Incalculable numbers of tiny creatures live in the spaces between the solid matter of crumbly soil and on the vast area of solid surfaces lining the spaces. The smaller spaces are occupied by water, while the larger spaces are filled with air. A researcher at Rothamsted, Britain's premier agricultural experimental station, once calculated that the surfaces lining all the tiny pores and passageways in a couple of tablespoons of soil add up to a quarter of a million square feet—the area occupied by a city block. No wonder there is room for so many creatures to live on the surfaces of soil granules and crumbs.

If the mineral particles of sand, silt, and clay were packed tightly together in the absence of other solid ingredients, pore spaces would remain among the particles, but the soil in these cases would be referred to as structureless. The smaller the mineral particles, the stickier the soil would be when wet and the harder it would be when dry. In such a tight, dense material, even the stronger roots of plants could not penetrate and air and water could barely circulate. Sticky, structureless clay soils are almost impermeable to water, while sandy soils without structure can lose water too rapidly.

It is amazing how a little organic matter added to the soil by living creatures or once living creatures can aggregate mineral particles and greatly improve the structure of a soil—and therefore the movement of water, air, and roots through it. But to maintain crumbs and granules in the soil requires constant addition of organic material and the constant activity of microbes. Fresh supplies of plant and animal material are needed to keep up with their constant decomposition by microbes. When organic material is partially decomposed and exists as long sinuous fibers that stretch from mineral particle to mineral particle, it is most effective at binding particles in granules and crumbs. But eventually even these organic molecules are broken down by microbes to elements of inorganic minerals as well as the simple compounds of carbon dioxide and water. This conversion of elements from their organic forms to their inorganic forms in soil is referred to as *mineralization*. Over time, soil that loses more plant and animal material than it gains eventually loses its crumbs, its structure, its pores, and its air spaces. Creatures that stir up the soil, that mix mineral matter with fresh supplies of organic matter, help keep the structure, the granules and crumbs, as well as the pores in a soil.

5. Diggers and Tillers of the Soil

This morning the ground beneath the oaks and maples of a neighborhood park is covered with many tiny piles of soil. Yesterday the same soil surface had been perfectly smooth, but during last night's rain, the neighborhood earthworms were busy. As they fed and burrowed, they were constantly swallowing soil, essentially eating their way through the ground and then piling their droppings as casts on the surface. Charles Darwin calculated that these casts can soon add up to 10 tons of soil brought to the surface in just one year on only an acre of land. That represents a substantial amount of earth moving for the earthworms that live on that acre.

Creatures that tunnel and burrow and dig participate in mixing the lower mineral-rich layers of the soil with the upper organic-rich layers. Their tunnels carry air and water to the deeper layers of the soil and provide channels along which roots often grow as they follow routes of least resistance. These burrowers inhabit and mold—mechanically and chemically—the uppermost portion of the Earth's soil, referred to as the *biomantle.* Many of the burrowing mammals can move massive amounts of soil—hundreds of pounds—when excavating a single den.

Spiders and insects, earthworms and crayfish may be much smaller than moles, gophers, and groundhogs, but they are certainly more abundant. When the earth-moving abilities of just one of these small creatures is multiplied by the number of these creatures that work the soil over time and over space, one realizes that collectively these creatures are capable of moving a phenomenal amount of soil.

Who are some of these diggers of the earth? Where do they dig, how much soil do they move, and how does their earth moving affect nearby plants and animals? To answer these questions, we should first examine the layers of soil through which these various diggers travel, the properties of the layers, and how mixing and redistributing these layers is important for the continual circulation of mineral nutrients in a soil. Without creatures living among its layers, a soil becomes static and its nutrients halt their circulation from one soil layer to another. Without the circulation of minerals, only a few rugged pioneers can establish a foothold on what has become a static, lifeless soil.

G. HOW PLANTS AND ANIMALS AFFECT THE LAYERS OF A SOIL

The mineral particles in a soil come mostly from rocks that have worn down over the ages; the organic particles arise from the decay of organisms that once lived in, on, or over the soil. The elements and compounds of a soil come from air, from water, from rocks, as well as from plants, animals, and other organisms. The various elements in a soil not only affect a variety of physical and chemical attributes of the soil, but they also influence which plants will take root in the soil and which animals will take up residence in it. Plants and

SOIL HORIZONS

Road cuts, trenches, and stream banks often expose the colorful and distinctive horizontal layers or *horizons* of a soil (fig. 21). These horizons tell about the elements and compounds that can be found in the soil. Over time, variations in the slope of the land as well as different climates, different plants, and different materials from which soils originate combine to give each soil a unique layering, from its surface down to its bedrock. Water that percolates down through a soil carries along with it elements and compounds from the soil's surface that settle at particular levels along the way. Often certain substances react with others that have already settled at another level in the soil. Wherever elements and compounds settle, combine, or react in the soil, they often leave colorful evidence of their whereabouts and give each soil its distinctively colored and textured layers.

Soil layers were first named after the first three letters in the alphabet: A, B, C. As studies of soils continued, other horizons were identified and their new names were added: O, E, R. Transitional horizons, such as AE, EB, and BC, are found between these six so-called master horizons (O, A, E, B, C, R). These master horizons and transitional horizons of the soil are often even further subdivided into thinner horizons according to their special properties of color or chemistry.

At the surface of the soil, the O (organic) horizon consists entirely of organic matter, some recently added to the earth and some well decayed. Beneath the O horizon lies the A horizon, often known as topsoil, where organic matter and mineral particles intermingle. Roots seem to grow best in the A horizon. In soils of grasslands, surface debris and the thick mat of decaying roots assure that the A horizon is thick and deep. In woodland soils, however, the A horizon is thinner than it is in grassland soils and is built up almost entirely by the decay of surface debris such as logs and fallen leaves. Relatively few tree roots decay each year to add their organic matter to the woodland soil. In grasslands, with their deep A horizons, most of the organic matter lies underground, whereas in a forest most of the organic matter is found aboveground in the living trees rather than in its shallow A horizon.

The E (eluviation, from Latin *ex* = out; *lavere* = to wash) horizon was once considered part of the A horizon. This layer loses most of its organic matter, clay, and nutrients from the leaching action of rain and melting snow as they percolate down through the horizons. The loss of matter and chemicals that give color to soil leaves little but pale sand in the E horizon.

The B horizon is where many elements, clay particles, and some organic matter accumulate as they are leached out of the overlying E horizon by rainfall or snow. Beyond the B horizon and beyond the reach of most plant roots

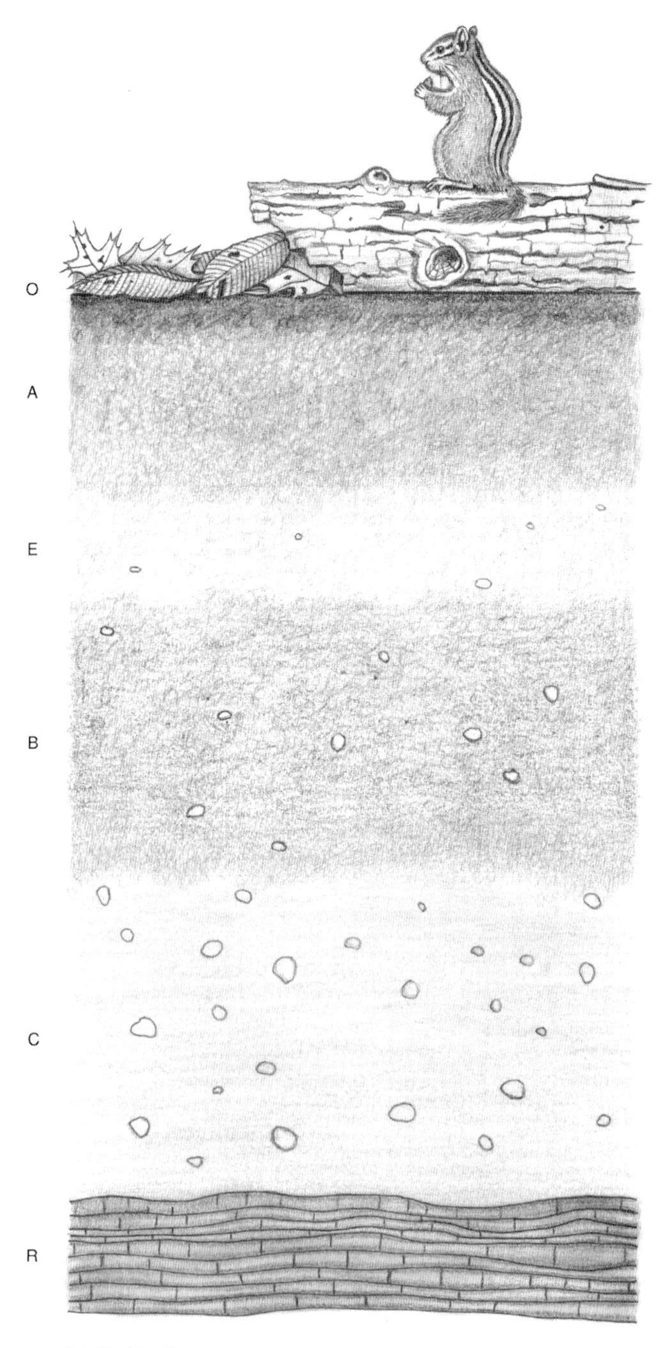

21. Soil horizons.

lies the C horizon, which is rarely perturbed by the activities of soil creatures nor any longer involved in the process of soil formation. This quiet horizon of the soil represents materials such as gravel and sand from which the soil above initially began to form.

Beneath all the other soil horizons, usually anywhere from 20 to 60 inches (50 to 150 centimeters) belowground, lies the R (rock) horizon, the layer of underlying bedrock. Rocks like limestone, sandstone, granite, schist, or slate may occupy this horizon. As long as the other horizons from O to C remain in place and uneroded, the R horizon remains solid and unaffected by any weathering and soil formation that take place in the horizons above.

---

animals that live in a particular soil are not only a consequence of that soil's properties—its texture, its water, its nutrients—but they are also frequently responsible for these properties. From observations made over many years in many places, we know that certain plants and animals are associated with certain soils. We can tell a lot about the layers of a soil from the plants and animals that live in it, and we can also tell a lot about who lives in a soil by examining the various layers of that soil.

For example, because there are few diggers and makers of burrows, little of the plant litter in a pine, hemlock, or other evergreen coniferous forest gets mixed with the deeper mineral soil. The soil shows a distinct layering, with the dark layer of plant litter clearly distinct from the pale, underlying mineral soil. In a forest of deciduous trees, on the other hand, there are many soil dwellers; their tillage results in well-mixed layers of soil that look very different from the distinct soil layers in a coniferous forest (fig. 22). The striking differences in the ways organic layers of soils blend with mineral layers in different types of forests was first noted over a century ago by the Danish forester P. E. Müller. In deciduous forests, the dark surface layer of plant debris, or the *mull* layer, blends imperceptibly with the brown mineral layer of the deep soil, and no obvious soil horizons can be seen. In a pine forest, the organic layer of plant debris, or the *mor* layer, remains separate and unmixed with the underlying mineral layers.

The surface layer of dark plant litter in a pine forest abruptly switches after a few centimeters to a white or gray mineral soil. These soils of coniferous forests are known as *podzols,* a Russian word mean-

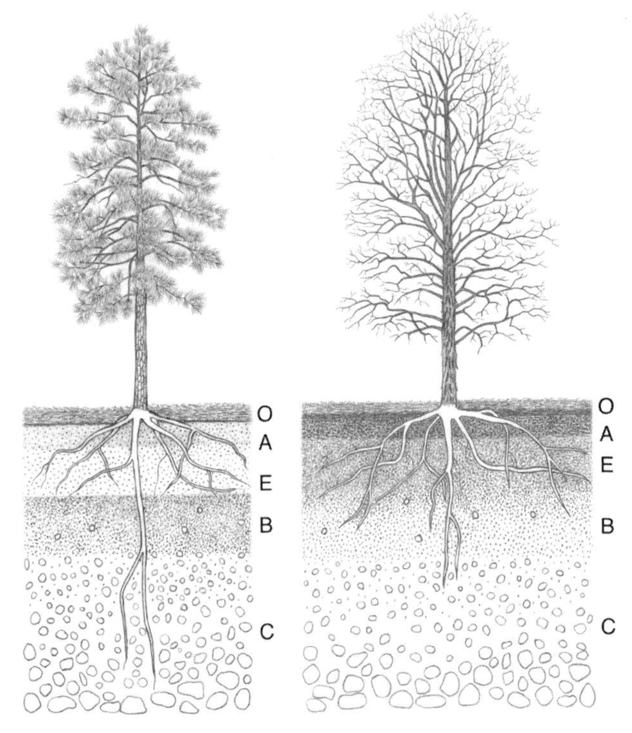

22. Under evergreen trees (left) the organic horizon of the soil is clearly a separate layer from the underlying mineral horizons, but in a deciduous forest (right), animals of the soil constantly mix the organic and mineral horizons.

ing "ashes underneath" and referring to the bleached appearance of the mineral soil that was once believed to represent the ashes left by forest fires. What has happened in the bleached layer or E layer of the podzol is that nutrients have been washed out of the layer by rain and melting snow. Almost the only materials left behind are the insoluble grains of sand. The light-colored sand gives the mineral soil its bleached appearance, and the nutrients that trickle downward come to lie in a red, brown, or black layer. Since the dark color of this soil layer comes from the toxic iron and aluminum ions that have settled in this horizon, tree roots in an evergreen forest tend to avoid these elements and grow best at the boundary between the sandy bleached layer and the toxic dark layer.

Obviously, it takes living organisms to keep various layers of soil in constant circulation, to keep essential nutrients from settling too deep in the soil, and to keep other toxic elements from concentrating and forming inhospitable layers in a soil. It also takes the decomposers to free these essential nutrients from dead plant and animal matter. Looking in on some of the many creatures that contribute their share to the circulation of nutrients in the soil, we see how they impart to soil its amazing complexity, its fertility, and its health.

PART TWO

Members of the Soil Community

A. MICROBES

Microbes are defined as those organisms that cannot be easily viewed without a microscope. These organisms easily outnumber all other organisms of the soil and include bacteria, actinomycetes, fungi, algae, and protozoa. In a square meter of soil the number of microbes often exceeds a trillion—a million million. They navigate the soil by swimming in water films and traveling in the spaces between soil particles. Bacteria and fungi act as decomposers and recyclers, while protozoa are the predators that keep populations of the other microbes in check. Algae and some bacteria are able to generate their own nutrients from the energy of light or chemical reactions.

Until recently, protozoa had been grouped with the invertebrate animals, but closer examination of their cellular and molecular structure has placed them in their own kingdom—Protozoa—a kingdom of one-celled organisms distinct from the Animal kingdom of multicellular organisms. Protozoa, along with other eukaryotic microbes, are also referred to as protists.

1. Eubacteria and Archaebacteria

For years scientists considered the decay and recycling of plant and animal debris in the soil to be a purely chemical reaction that did not involve any living creatures. However, the clever experiments of the French biologist Louis Pasteur in the latter half of the nineteenth century established beyond question the importance of bacteria in

Kingdom Archaebacteria (archaebacteria)
Kingdom Eubacteria (true bacteria)
Place in food web: decomposers, bacterial partners of plants, producers (autotrophic bacteria)
Impact on gardens: allies
Size: 1.0–10 μm (0.001–0.01 mm)
Number of species described: 279 (Archaebacteria), 6,501 (Eubacteria)

decomposing organic matter and liberating its nutrient elements in forms that plants can use. Bacteria decompose not only naturally occurring organic compounds, derived from living or once living organisms, but also human-made compounds like pesticides. After all, there are plenty of bacteria to handle decomposition—a billion of them in each gram of dry, fertile soil. If all of the bacteria on a single acre of soil were weighed, they would tip the balance at one and a half tons (1.2 metric tons).

The number of different species of bacteria in a given volume of soil has not been easy to estimate, for they have few distinguishing characters that are visible to the eye. They have more or less the same size, shape, and overall appearance. Although they look the same on the outside, their genetic makeup can be very different. From an analysis of genes and the deoxyribonucleic acid (DNA) that makes up these genes, two Norwegian biologists in 1990 set out to estimate the number of bacterial DNAs in a sample of forest soil that were at least 30 percent different from one another.

For some larger animals such as birds and mammals the genetic differences separating species are often far less than 30 percent. The genetic difference between humans and chimpanzees, for example, is only 2 or 3 percent. From their analysis of the DNA extracted from all the bacteria in one gram of woodland soil, the Norwegian biologists concluded that anywhere from 4,000 to 5,000 species of bacteria live in that one little parcel of the earth's surface. More recent, and probably more accurate, calculations published in a 2005 issue of *Science* by researchers at Los Alamos National Laboratory in New Mexico estimate that the number of distinct species of bacteria in a gram of pristine soil is actually closer to one million.

Bacteria seem to inhabit more areas of the soil than any other creatures; scientists have found bacteria dwelling as far below the surface of the soil as 3.7 kilometers (slightly over two miles). Imagine how many different species of bacteria—with many different talents—inhabit the different soils of our planet if almost a million species inhabit less than a tablespoon of it.

All bacteria are small with a limited number of forms. Based on their studies of bacterial shapes and sizes alone, biologists had no idea just how many different species of bacteria there are or how different some of these bacteria could be until the advent of genetic analysis. Comparison of different bacterial molecules revealed that a large group of bacteria that live in extreme environments are actually more similar to plants, animals, and fungi than they are to the other bacteria. These bacteria are often referred to as "extremophiles" that inhabit hyperthermal environments such as hot springs, hypersaline environments such as salt-encrusted soils, and anaerobic environments such as water-logged soils or the digestive tracts of soil arthropods. Other bacteria are so different from these extremophiles that they are considered members of a separate kingdom as well as a separate domain of life, the Eubacteria. The extremophiles of the microbial world have been placed in the domain and kingdom Archaebacteria or simply Archaea. They were considered rare in most soils until recently, when they were discovered in abundance around the roots of garden plants, in an environment obviously considered far less extreme than hot springs or salty soils (figs. 23–24).

Bacteria of the soil are more abundant around roots—in a zone known as the rhizosphere (*rhizo* = root; *sphero* = sphere)—than they are anywhere else in the soil. The number of bacteria that inhabit the rhizosphere is 10, 20, or even several hundred times as great as the number of bacteria in areas of the soil that are rootless. Roots are constantly releasing organic substances, including dead cells, that can be broken down and used as a source of energy by certain bacteria. No doubt the plant benefits from the proximity of these numerous bacteria and their talents for enriching the soil.

Plants are dependent on bacteria to ensure that the essential elements of soil are available to their roots. Two types of bacteria provide nutrients to the roots from different sources: heterotrophic ("other-nourishing") and autotrophic ("self-nourishing"). Heterotrophic bacteria depend on organic substances in the soil for their nourishment and transform this organic matter into plant nutrients. Autotrophic bacteria can generate their own organic matter from carbon dioxide and in addition can transform inorganic matter and minerals

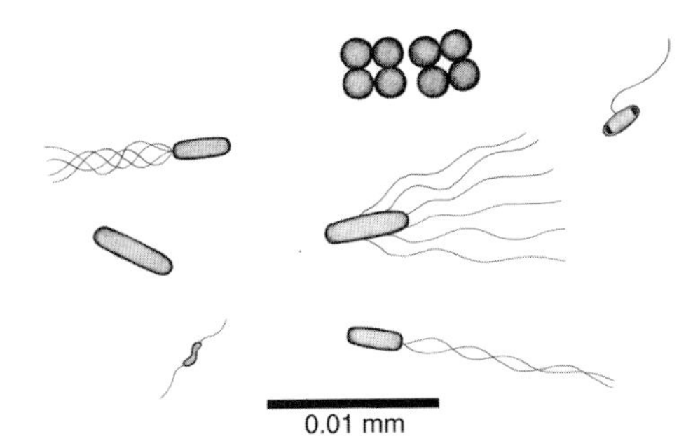

23. Bacteria of the soil come in a variety of sizes as well as shapes. They carry out jobs involving chemical transformations that are essential for the survival of all life on earth. The bacteria shown here are (clockwise from upper left): *Pseudomonas, Sarcina, Desulfuromonas, Azotobacter, Rhizobium, Thiomicrospira,* and *Clostridium.*

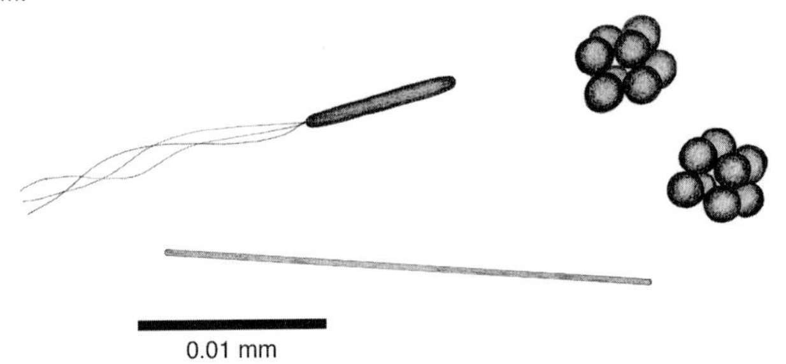

24. Here are some of the Archaebacteria that are found in soils with "extreme" properties. In these extreme environments they dwell with little, if any, competition from other organisms. In shape and size, they can easily be mistaken for microbes in the kingdom Eubacteria. *Halobacterium* (*halo* = salt; *bacterium* = rod) of salty soils imparts a distinctive pink color to salt crusts (top left). *Methanosarcina* (*methano* = methane; *sarcina* = bundle) lives in strictly anaerobic environments and is one of many methane-producing Archaebacteria that convert carbon dioxide or other carbon-containing compounds in their environments to methane (top right). *Thermoproteus* (*thermo* = heat; *proteus* = changing form) thrives in hot springs where temperatures are only a few degrees below the boiling point of water (bottom).

into nutrients that plants can use. Whatever substances bacteria convert into nutrients for plants, all bacteria ultimately provide elements in forms that plant roots can readily absorb and use for their growth. Bacteria oversee a vast range of chemical transformations in the soil. All of the 18 essential elements used first by plants then by animals are continually recycled, thanks to the bacteria of soil.

The billions upon billions of soil bacteria exert an extraordinary influence on the health of the earth. Acting as crucial recyclers and decomposers of plant and animal remains, heterotrophic bacteria control the amount of organic matter and organic pollutants remaining in the soil as well as the levels of carbon dioxide released from soil into our atmosphere. Ninety percent of the carbon dioxide produced by life on earth arises from the activities of bacteria and their fellow microbes, the fungi. The metabolic activity of the microbes in the top six inches (15 centimeters) of a single acre of rich soil easily surpasses the metabolic activity of 50,000 human beings. In the final stages of decomposition of organic matter in soils, heterotrophic bacteria liberate not only carbon dioxide but also elements essential for the survival of all plant and animal life on earth.

Autotrophic bacteria convert many inorganic compounds that contain essential but inaccessible elements for plant growth to elements that plants can avidly consume. Like green plants, autotrophic bacteria consume carbon dioxide, thus diminishing the levels of this gas that has been implicated in global warming.

These simple creatures, the bacteria, have managed to carry out transformations of elements from one inorganic form to another inorganic form, from organic to inorganic, from living organisms to minerals—transformations that no plants or animals have managed to achieve during their history on earth. It is to the bacteria of the soil, with a little help from the larger inhabitants of the soil, that most of the credit for the constant renewal of our earth is due.

2. Actinomycetes

Kingdom Eubacteria (also known as Bacteria)
Place in food web: decomposers, bacterial partners of plants
Impact on gardens: allies
Size: filaments 0.05–2.0 µm (0.00005–0.002 mm) in diameter
Estimated number of species: 1,460

The smell of good earth that rises from a newly plowed field or from digging in the rich, dark soil of a hardwood forest is the odor given off by the millions of actinomycetes that live in the soil. This earthy odor comes from a compound appropriately known as geosmin (*geo* = earth; *osmi* = odor) that arises as actinomycetes break down or-

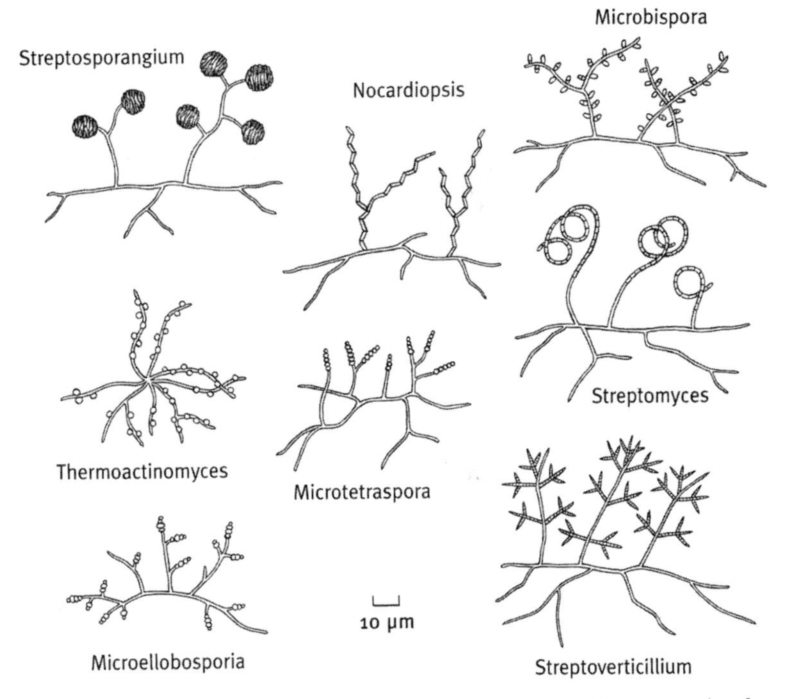

25. Actinomycetes are special bacteria that produce spores and form networks of filaments.

ganic matter. Actinomycetes resemble both fungi and bacteria. They form fungal-like colonies in the soil with their filaments radiating outward, whence the name (*actino* = ray; *mycos* = fungus). Like fungi, actinomycetes form networks of filaments in the soil (fig. 25). However, the spores from which the filaments sprout look more like bacterial spores than fungal spores. Even after being studied for many decades, actinomycetes still puzzled those who tried to pigeonhole them. Did they represent a common ancestor of the fungi and bacteria, or were they somewhere between the two groups? Although on the outside actinomycetes can take on forms that resemble both bacteria and fungi, on the inside their molecules clearly look more like those of bacteria than those of fungi.

Actinomycetes help in the decomposition of soil organic matter, but they are better known as sources of antibiotics. In the soil, antibiotics like actinomycin, tetracycline, neomycin, and candicidin help maintain checks and balances on burgeoning bacterial populations.

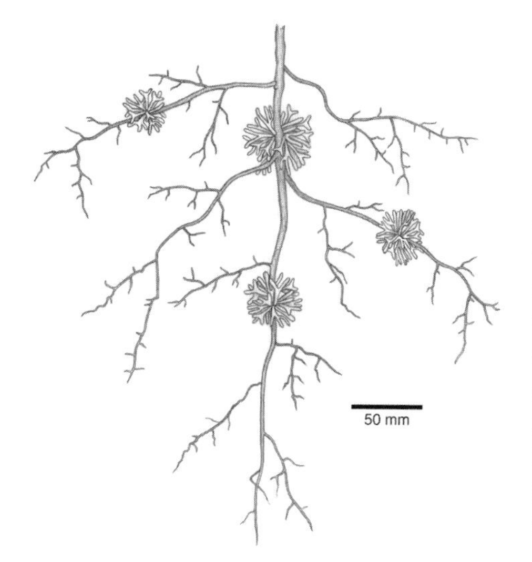

26. Some actinomycetes associate with roots of woody shrubs and form nitrogen-fixing nodules where dinitrogen gas in the air is converted into usable nitrogen compounds. These nodules formed on roots of the prairie plant, New Jersey tea.

One of the better-known antibiotics, streptomycin, is produced by one of the more common actinomycetes by the name of *Streptomyces*. The antibiotics of actinomycetes had been controlling soil bacteria for millions of years before we discovered their usefulness in controlling our own disease-causing bacteria.

Actinomycetes are among the few soil organisms that have the ability to take dinitrogen gas (N_2) from the air and convert it to ammonia (NH_3) in the soil. Unlike the rhizobial bacteria that associate only with the roots of legumes like peas, beans, and clovers, actinomycetes associate with plants that are not legumes (fig. 26). The actinomycetes invade root hairs of these plants just as rhizobia invade root hairs of legumes. Also like rhizobia, actinomycete filaments penetrate through the root cells and induce root nodules; but unlike the typically small, round nodules formed by rhizobia, nodules formed by actinomycetes are knobby and branched and sometimes several centimeters in diameter.

New plants that bear root nodules with nitrogen-fixing actinomycetes are continually being found, and their contribution to worldwide nitrogen fixation was not appreciated until the latter half of the twentieth century. Symbiotic rhizobia, actinomycetes, and blue-green algae as well as several free-living microbes fix about 140 million metric tons of nitrogen each year—about twice as much as the supply of nitrogen fertilizer produced by industrial sources.

3. Algae

Kingdom Eubacteria (blue-green algae or cyanobacteria)
Kingdom Plant (green and red algae)
Kingdom Chromista (algae with flagella, yellow-green algae, golden algae, diatoms)
Place in food web: producers
Size: filaments or cells 2.0–25 µm (0.002–0.025 mm) in diameter
Impact on gardens: allies
Estimated number of species: 7,000 green algae; 1,500 blue-green algae; 4,000 red algae; 6,000 diatoms, golden algae, and yellow-green algae; 800 algae with flagella.

Algae are usually associated with ponds, streams, marshes, and mud puddles, but closer looks at the algae that live in soil have changed this conventional view of them. Algae are even common in the soils of deserts and have been revived from soils that have been stored dry for as long as 83 years. While they are in this dry, dormant state, algae can endure temperatures exceeding the boiling point of water at 100°C (212°F) as well as temperatures down to −195°C (−320°F). Here in the soil is where algae and fungi probably first discovered that they could establish compatible and fruitful partnerships in the form of lichens.

Like the lichens that arose from them, algae can be hardy pioneers and builders of soil (figs. 27–28). They were the first colonists on the Indonesian island of Krakatoa after its violent eruption in 1883. Like other green plants, algae have the ability to capture the sun's energy and produce their own sugars and organic matter. As they use the energy they produce from photosynthesis, they give off carbon dioxide like all other living creatures. The corrosive carbonic acid that forms when carbon dioxide combines with water contributes to soil formation by helping to digest any rocks that might be in the immediate vicinity. As many as a hundred million algae have been counted per gram of soil. All these algae add a phenomenal amount of organic matter to the soil. In some Arizona soils, algae annually add about six tons of organic matter to the top three inches of each acre. Blue-green algae, like certain bacteria and actinomycetes, can fix nitrogen from the atmosphere and add this essential nutrient to impoverished soil, once again pioneering the way for the plants that will follow.

The organic matter that algae add to the soil is often sticky and helps improve the structure of soil by holding clay and humus par-

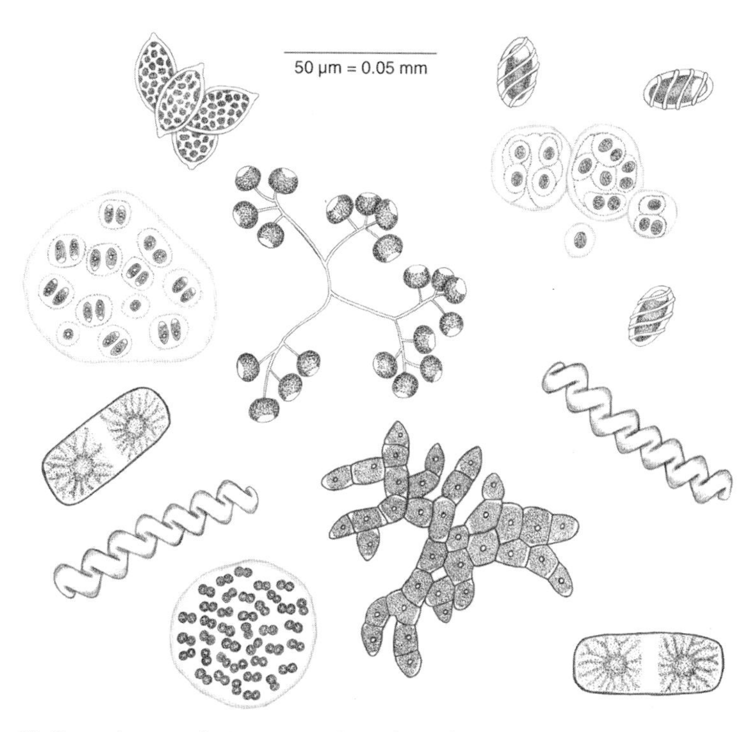

27. Some algae are plants, some are bacteria, and some are protists. They all use the energy of sunlight to produce their own nutrients. A variety of soil algae are shown here (clockwise from center): *Dictyosphaerium, Gleocapsa, Scotiella* (3 cells), *Spirulina, Cylindrocystis, Protoderma, Aphanocapsa, Spirulina, Cylindrocystis, Palmella, Oocystis.*

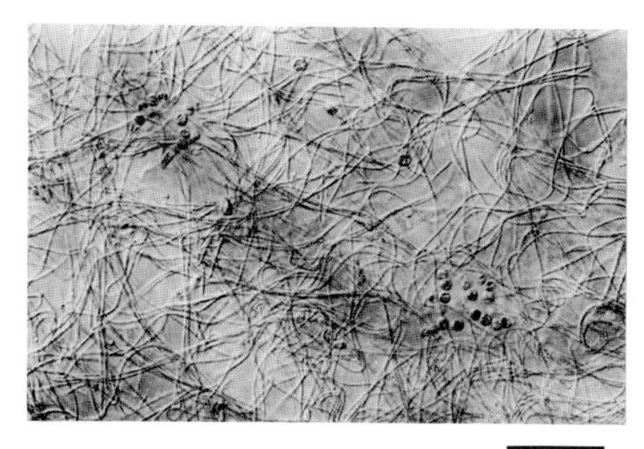

28. A view of an algal jungle that has formed a crust on nutrient-poor soil.

ticles together in loose clumps. When soil structure is improved, the soil is able to hold more water in its pore spaces and erosion is markedly reduced. The soil of bare, eroded fields that has been colonized by soil algae contains about 10 percent more water than comparable soil without algae.

As producers of organic matter from the energy of sunlight, algae give more to the soil community than they take. Algae nurture the soil with organic matter and nitrogen that they leave behind; they also provide nutritious feasts for the many small arthropods that feed on them.

4. Fungi

Kingdom Fungi
Phylum Zygomycota (sugar fungi, molds, Trichomycetes)
Phylum Ascomycota (yeasts, sac fungi, and some fungi that form mushrooms)
Phylum Basidiomycota (fungi that produce many forms such as mushrooms, shelf fungi, jelly fungi, puffballs, rusts, and smuts)

Place in food web: decomposers, fungal partners of plants and some arthropods, predators
Impact on gardens: allies/adversaries
Size: filaments about 3.0 µm (0.003 mm) in diameter
Number of described species: 75,000, of which about 700 have been described from soil.

Along with bacteria, fungi are the main recyclers of nutrients and a major source of nutrition for many of the animals of the soil. But fungi are versatile and play many other roles in the soil community. In addition to certain fungi forming mycorrhizal partnerships with green plants as they have for at least 400 million years, other fungi known as Trichomycetes (*tricho* = hair; *myco* = fungi) form partnerships with a number of soil-dwelling arthropods such as millipedes (see figs. 65–67, plates 16–17). These specialized fungi live only in association with specific arthropods and only in the guts of these arthropods. Presumably both the fungus and the arthropod benefit from this association. Some soil fungi protect plants by eating nematodes (see fig. 39) and insects before these animals have a chance to feed on roots. Many fungi have established these mutually beneficial relationships with members of other kingdoms, with only a minority of fungi infecting and harming animals and plants. Because only about

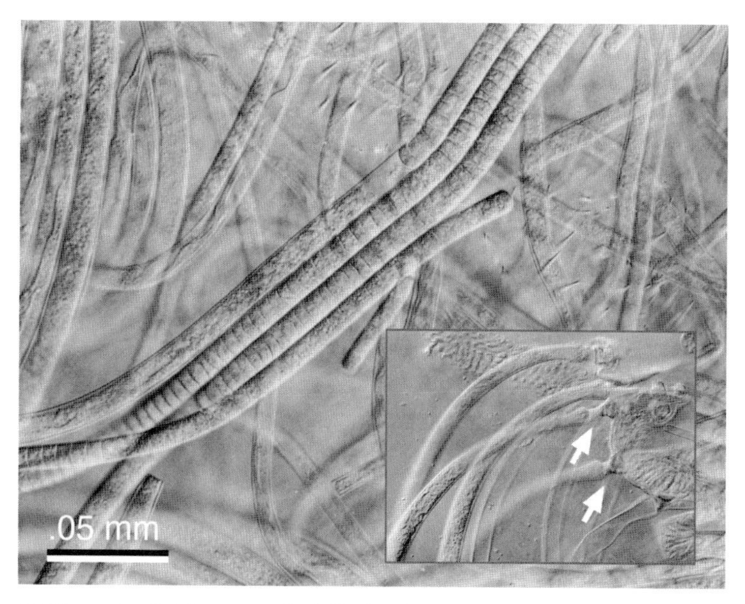

29. Filaments of trichomycetes look like long hairs growing on the lining of a millipede's hindgut. The inset shows that each filament has a special holdfast attaching it to the cuticle that lines the gut (arrows).

5 percent of the estimated 1.5 million species of fungi have been described, we can anticipate discovering new species as well as new fungal partnerships for years to come (fig. 29).

Plant debris that falls to the ground is colonized by a succession of fungi. As fresh plant matter gradually decomposes, the simplest compounds of the dead plant matter are broken down first by a group of fungi known as sugar fungi. These first colonizers of leaf litter—the molds of the soil—obtain their energy for survival from simple sugars. They can digest simple sugars, but not the tough fibers of plant tissues made up of chains of sugars known as cellulose and hemicellulose or the lignins that hold these sugar chains together. To deal with possible competition for the simple sugars in the leaf litter, these pioneering colonizers of newly fallen plant debris grow very quickly and are very abundant in the soil; in addition they secrete antibiotics that discourage growth of nearby bacteria and fungi. Sugar fungi are also among the few fungi that can survive the high temperatures that they often generate as they extract the first nutrients from fresh plant debris. In a compost pile these sugar fungi begin the decomposition by releasing a burst of energy that can raise the temperatures in the center of the compost pile to 65–75° C (150–165° F).

After these sugar fungi have exhausted their food resources of simple sugars and begin to die, the next stage of colonization begins as other fungi move in to begin devouring the tough fibers that remain in the partially decomposed plant litter. The filaments of sac fungi and fungi that form mushrooms take the place of the sugar fungi. These fungi are covered with a layer of material that lacks cellulose and resembles in some ways the cuticle of insects. Not having cellulose in their own cells, they can set about digesting the cellulose and hemicellulose of plant litter with impunity. After these fungi have completed their digestion of wood, they leave behind a brown, crumbly residue rich in lignin. Their handiwork has earned them the name of brown rot fungi. Another group of fungi, made up mostly of mushroom producers, not only break down hemicellulose and cellulose but they also decompose lignin, the most resistant of the plant residues. The bleached and stringy appearance of wood that has been decayed by these mushroom formers has earned them the name of white rot fungi (plate 12).

Fungi are specialized not only for decomposing plant remains but also for decomposing animal remains and animal dung. Hair and hooves, claws and feathers are food for certain sac fungi, and many molds grow on the droppings of animals.

Whether several fungal filaments or hyphae that are recovered from the soil are part of just one fungus or are part of different fungi is by no means easy to ascertain without laboratory testing. Estimates of the abundance of fungi in the soil are usually made in terms of lengths of fungal filaments rather than in terms of numbers of individuals, with each descended from a single spore. So when biologists actually started looking carefully at the total length of hyphae in a single soil fungus, they were rather astounded by the lengths to which fungal hyphae can grow and the acreage over which a single fungus can roam. In 1992, a 38-acre fungus was reported from the woods of Iron County, Michigan; soon thereafter a fungus occupying 1,500 acres was reported living in a forest of the Pacific Northwest. By some measures, these fungi are the largest living creatures on earth. Fungi this large are at least 1,000 and maybe even 10,000 years old.

Soil fungi are easy to observe in the field and in the laboratory.

30. The dark rhizomorphs of bootlace fungus form intertwining strands on the surface of this decaying elm log.

Damp, rainy weather encourages the growth of fungi, so not surprisingly fungi that produce mushrooms are most common after rainy weather in spring, late summer, and fall. Many of these have mutually beneficial associations with roots of nearby trees. Fungal filaments sometimes aggregate into strands of several thousand individual filaments. These strands look and act like growing plant roots, forcing their way into hard, dead wood and accelerating its decomposition. These rhizomorphs (*rhizo* = root; *morph* = form) not only look like roots, but they also look even more like boot laces, and the dark rhizomorph of the tasty autumn honey mushroom (genus *Armillaria*) is known as bootlace fungus (fig. 30).

Rhizomorphs and hyphae can be found on fallen logs, under their bark, or throughout the rotting wood about any time of year, even on hot, dry days. Leaves of the forest floor are coated with hyphae that form a variety of delicate, intertwined patterns. Many fungi that happen to be present on a small fragment of wood, fallen leaf, or soil can be enticed to grow by placing the fragment on a moist piece of sterile filter paper in a clean, covered container. By examining them under a microscope, a person can gain a firsthand and close-up appreciation of how fungi and their soil companions are able to recycle the roughly ton and a half of leaves that fall to the ground each autumn on every acre of deciduous forest (fig. 31).

31. Hyphae of the many species of fungi that colonize and decompose the litter of the forest floor do not form mushrooms but produce asexual spores or conidia (*coni* = dust; *idia* = little) in special structures called conidiophores (*conidia* = spores; *phore* = carry). Conidiophores of several species with a variety of distinctive forms are shown here.

5. Chytrids, Hyphochytrids, and Oomycetes

Kingdom Fungus
Phylum Chytridiomycota (chytrids)
Place in food web: parasites of plants, fungi, and animals; decomposers
Impact on gardens: allies/adversaries
Size: 10–20 µm (spore diameter)
Estimated number of species: 1,000

Kingdom Chromista
Phylum Hyphochytriomycota (hyphochytrids)
Place in food web: parasites of fungi and algae, decomposers
Impact on gardens: allies
Size: 10–20 µm (spore diameter)
Estimated number of species: 23

Phylum Oomycota (oomycetes)
Place in food web: parasites of plants, fungi, and invertebrates; decomposers
Impact on gardens: allies/adversaries
Size: 3–5 µm (filament diameter)
Estimated number of species: 600

Chytrids (*chytro* = pot) and hyphochytrids (*hypho* = web; *chytro* = pot) are small and inconspicuous organisms of the soil, but they are common and widely distributed on earth. Both of these organisms are extremely small and are usually observed with microscopic examination of the plants, animals, or organic debris that they colonize. Although these two groups of organisms both have cells with flagella and were once considered to be closely related, they are now considered distantly related and members of different kingdoms because

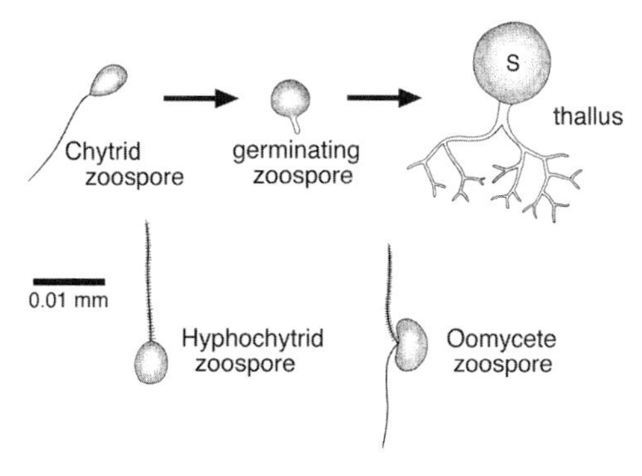

32. Chytrids are the only true fungi with swimming cells called zoospores. Each zoospore has a single posteriorly directed whiplash flagellum. When a spore germinates, it sprouts a thallus (*thallus* = young shoot) that grows without forming any cross walls until the thallus produces a reproductive structure (sporangium, s).

Hyphochytrids and oomycetes are not true fungi. Each hyphochytrid zoospore has a unique tinsel flagellum that is directed forward. Each oomycete zoospore, in addition to having an anterior tinsel flagellum, has a second whiplash flagellum that is directed backward.

of the striking differences in their flagella. Chytrids are the only true fungi that have flagella. Flagella are very conserved structures and are expected to differ only in organisms whose lineages are separated by at least a hundred million years. All hyphochytrid cells also have a single flagellum that is directed forward and covered with fine lateral filaments (tinsel flagellum); practically all chytrid cells have a single, smooth (whiplash) flagellum that is directed backward.

As members of the soil food web, species in both kingdoms contribute to the decomposition of some of the more persistent organic debris such as shed snake skins, insect cuticles, pollen, and the keratin protein of fur and antlers. One hyphochytrid is a parasite of oomycetes, which, in turn, are parasites on plants of economic importance. With many more species in their phylum, the chytrids have specialized as parasites on many more groups of organisms: plants, invertebrates, fungi, and even other chytrids. One chytrid is known to infect amphibians that live underground: salamanders and caecilians.

Members of the phylum Oomycota are important decomposers in aquatic habitats and moist soils; but they are better known as water molds in aquariums and as pathogens of plants responsible for damping-off diseases, potato blight, downy mildew, and sudden oak death. The soil-inhabiting members of the Oomycota can attack roots, seeds, fungi, and invertebrates in their environment. Oomycetes are set apart from the chytrid fungi and hyphochytrids by having zoospores with two flagella, not just one. Each oomycete spore has a tinsel flagellum directed forward and a shorter whiplash flagellum directed backward (fig. 32).

6. Lichens

A mix of several kingdoms
Kingdom Eubacteria (blue-green algae or cyanobacteria)
Kingdom Chromista (golden and brown algae)
Kingdom Plant (green algae)
Kingdom Fungus (98% Ascomycota, 2% Basidiomycota)

Place in food web: decomposers, producers
Impact on gardens: allies/absent
Size: algae, 2–25 µm in filament or cell diameter
fungi, 3–10 µm in filament diameter
Estimated number of species: 18,000

Many of the soil algae and soil fungi gave up their free-living existences to become members of a very successful alliance that draws on the strengths of each partner. Fungi are heterotrophs that survive on organic molecules and absorb inorganic nutrients, while algae are autotrophs that can capture the energy of sunlight and manufacture their own organic nutrients. The algal member of the partnership provides organic nutrients, and the fungal filaments that embrace the algae absorb essential inorganic nutrients from rock or soil.

Usually one, but sometimes two, species of algae join with a specific fungus to form a lichen that can survive harsher conditions than can either the algal or the fungal partner on its own. Lichens have an uncanny ability to extract minuscule quantities of nutrients from the poorest of soils and the hardest of rocks (plate 13). Even subfreezing temperatures, scorching sunlight, and long droughts do not trouble the steadfast alliance of algae and fungi; the lichen partnership endures the most adverse conditions. Only in the presence of unlimited nutrients and plenty of moisture, when the living is easy, may algal cells of a lichen grow as free-living algae, abandoning their fungal partners.

Another attribute arising from this partnership or symbiosis (*sym* = together; *bio* = life; *-sis* = the act of) of algae and fungi is the production of the unusual acids of lichens. On their own, neither algae nor fungi produce the hundred or so organic acids that are unique to lichens. In addition to corroding rocks, the acids of lichens readily combine with elements from rocks and enhance the solubility of these elements in the soil. These acids also inhibit the growth of other microbes and show antibiotic activity. Since lichens grow extremely slowly, expanding at the rate of one millimeter or less each year, a little antibiotic activity can protect them from encroachment by more rapidly growing microbes that share their rock or patch of soil.

Lichens can also warn of pollutants in the environment. Their ability to scavenge and concentrate trace amounts of elements from the environment—even toxic ones—has had disastrous consequences for lichen populations in polluted places. Their disappearance from a landscape is a warning to the rest of us.

7. Slime Molds

Kingdom Protozoa

Phylum Myxomycota (acellular slime molds)
Place in food web: predators of bacteria and protozoa, decomposers
Impact on gardens: allies
Size: 5.0 µm or 0.005 mm (diameter of cell)–10.0 mm (height of fruiting body)
Estimated number of species: 700

Phylum Acrasiomycota (cellular slime molds)
Place in food web: predators of bacteria and protozoa, decomposers
Impact on gardens: allies
Size: 5.0 µm or 0.005 mm (diameter of cell)–10.0 mm (height of fruiting body)
Estimated number of species: 65

The creatures known as slime molds that creep through decaying wood and between rotting leaves seem to be part fungus, part protozoa. They have fruiting bodies like fungi at certain times in their life cycles, but they move and act like protozoa at other times. Like amoebae, they creep and spread over logs, twigs, and fallen leaves leaving a trail of slime in their wake (plate 14). As they glide along they engulf bacteria, some protozoa, as well as small pieces of decaying plant remains. Then, within only a few hours, these blobs of undifferentiated tissue can transform into mushroomlike masses of often richly colored and elaborately sculptured fruiting bodies (fig. 33).

Not surprisingly, biologists were never too sure where in the tree

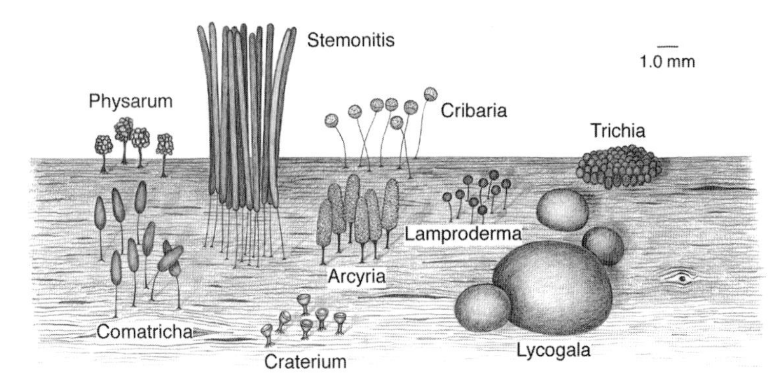

33. The fruiting bodies or sporangia of slime molds often have unusual and colorful forms. The genera of different slime molds appear here.

of life to place these creatures. Some considered slime molds fungi while others placed them among the protozoa. Slime molds are now considered members of the same kingdom to which protozoa belong, the kingdom Protozoa; but within this kingdom, they are still considered unique enough to be given two phyla of their very own.

Even among the slime molds two clearly distinct groups—the acellular slime molds and the cellular slime molds—are found in the soil and on decaying leaves and wood. Each acellular slime mold grows and moves as a single immense amoeba known as a plasmodium, often many inches across and usually bright yellow. The cellular slime molds, on the other hand, grow by dividing into many little independent amoebae.

When the surroundings of slime mold amoebae and plasmodia become drier and brighter, and their food becomes scarcer, they prepare to settle into the sedentary life of a fruiting body. The multitude of individual amoebae gathers together to form a slug-shaped aggregate. As the tiny slug imperceptibly creeps over the forest floor, individual amoebae within the aggregate sacrifice their independence to interact in unison and harmony like the individuals in a school of fish or a flock of geese. The single large plasmodium of each acellular slime mold turns into a cluster of fruiting bodies, whereas the aggregate of many small amoebae that make up each slug of a cellular slime mold turns into a single fruiting body.

Slime molds are everywhere, but unless you set out to find them you will probably never see them. On a walk through the woods in summer or fall, when logs and leaf litter are moist from earlier rains, you are very likely to spot small patches of what can best be described as bright yellow, lacy jelly spread thinly over the surfaces of rotting logs, twigs, or leaves. At drier times during the year, you are more likely to spot clusters of colorful fruiting bodies of the slime molds on the same surfaces.

If you spot a yellow plasmodium on a well-decayed log, you can try cultivating it at home. Place a small fragment of the plasmodium attached to the underlying log in a loosely capped jar containing a piece of absorbent paper, such as a paper towel. Keep the paper moist, but not wet, and feed the plasmodium some uncooked rolled oats every few days. Also keep the jar in the dark and only expose it to light when you briefly examine the slime mold. The plasmodia grow at phenomenal rates. Like a giant amoeba, the organism will take solid food into its jellylike body and may cover every surface of the jar by the next day.

8. Protozoa

Kingdom Protozoa
Phyla Amoebozoa and Cercozoa (amoebae)
Phylum Ciliophora (ciliated protozoa)
Phylum Heliozoa (heliozoa)
Phyla Euglenozoa and Metamonada (flagellates)

Place in food web: predators of bacteria
Impact on gardens: allies
Size: 0.005–0.16mm
Estimated number of species: 65,000

At least 300 species of protozoa are known to live in the thin film of water that usually lines the innumerable pores of the soil. If the water film that lines the pores in two tablespoons of soil were spread out flat, it would cover the area of a whole city block, or about 24,000 square meters. Quite a few protozoa can squeeze onto a city block, so one European estimate of 10 billion protozoa living in the top six inches (15 centimeters) of a square meter of meadow should not seem outlandish.

Soil protozoa come in four main forms: those that have flagella (*flagellum* = whip); those that have cilia (*cilia* = small hair); those that are amoeboid (*amoeba* = change) and constantly changing their shapes can either have a protective shell called a test (*testa* = shell)

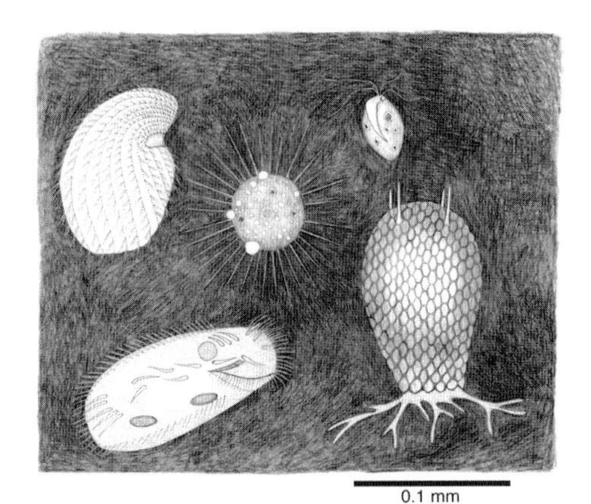

34. Protozoa swim in the film of water that covers soil particles. The protozoa shown here are (clockwise from center): heliozoa *Actinophrys*, flagellate *Tetramitus*, testacea *Euglypha*, ciliate *Oxytricha*, ciliate *Colpoda*.

or no shell at all (fig. 34). Protozoa have definite preferences for certain soils. Shell-bearing amoebae, or testacea, are more common in forests than they are in cultivated fields. Whereas about 100 testacea may live in a teaspoon of cultivated soil or desert soil, about a thousand times this number can be found in the litter and soil of forests. Amoebae without shells but with fine, radiating rodlike pseudopods (*pseudo* = false; *pod* = feet) are more common in wet soils. Their graceful, starlike forms have suggested the name of heliozoa (*helio* = sun; *zoa* = animals). The smaller amoeboid protozoa and protozoa with flagella seem to be the more common forms in most soils; and members of the genus *Colpoda* seem to be the most common of those protozoa with cilia.

Soils can often dry out, and water films can evaporate from pores of the soil. When this happens, soil protozoa lose their normal shapes and enter an immobile and inactive state called a cyst. There they wait until damper days arrive. In some soils and at particular times of year, most, if not all, of the protozoa may exist in this state of suspended animation.

Although larger protozoa, such as testacea, sometimes feed on algae, fungi, and even plant debris, most protozoa feed on bacteria; their major influence on cycling of organic matter and on humus formation is through their control of bacterial populations. In fact, protozoa seem to be the principal predators of soil bacteria. By one estimate 90 percent of the bacteria consumed in the soil are eaten by protozoa; the other 10 percent are eaten by small worms called nematodes, most of which are no more than a millimeter long. Larger nematodes, in turn, eat many of the protozoa that have fattened on bacteria.

New species of flagellated protozoa are continually being discovered in the digestive tracts of soil animals. These protozoa were first described from the hindguts of termites. The protozoa supposedly have established mutually beneficial relationships with their animal hosts and are part of the relatively unexplored microbial worlds found inside animals of the soil.

Animal Kingdom

This kingdom is dominated in numbers by the invertebrates; but in terms of size, the animals are clearly dominated by mammals. Some of the smaller animals, such as rotifers and tardigrades, are small enough to be considered microbes, but conventionally the term microbe excludes animals, all of which are multicellular. Unlike plants and some microbes, animals are heterotrophs and depend on other organisms—either alive or dead—for their nourishment. The major division within the Animal kingdom is drawn on the basis of whether an animal has a backbone or not.

B. INVERTEBRATES: ANIMALS WITHOUT BACKBONES

About 97 percent of the species in the Animal kingdom are animals without backbones and internal skeletons. They inhabit the land, the sea, and—in the case of the insects—the air. In the soil, invertebrates range in size from the giant three-meter earthworms of Australia that dig burrows deep into the ground to the microscopic rotifers and tardigrades that move about in the thin films of water that coat particles of sand and silt. The number of soil invertebrates often exceeds several million in a square meter of soil. They can be predator or prey, herbivores or fungivores; as scavengers and shredders of plant and animal matter, they facilitate the recycling work of the microbes.

Animals without Backbones or Jointed Legs

1. Flatworms

Flatworms are creatures of the night, the time when they can glide from beneath rocks, logs, and leaf litter with little danger of drying up

Phylum Platyhelminthes (flatworms)
Class Turbellaria (free-living flatworms)
Place in food web: predators
Impact on gardens: adversaries
Size: 0.2–600 mm
Estimated number of species: 3,000

during their travels. In addition to having nocturnal habits, our native American flatworms are colored like their surroundings and are tiny, ranging in size from 0.2 mm to 70 mm. They have a well-deserved reputation for being difficult to find. Even though some tropical flatworms can grow to 600 mm in length and often come with stripes and bright colors, they still have very simple body plans.

Not only do all flatworms survive without a circulatory or respiratory system, but they also have a gut with only one opening that serves as both mouth and anus (fig. 35). The flatworm's two-way gut has so many branches and ramifications throughout its body that whatever nutrients it obtains from digestion are quickly passed to every tissue of the body. Each flatworm also has a brain, a nervous system, and specialized sensory cells including two eyes. By being so flat, a flatworm has a large surface area for inhaling oxygen and exhaling carbon dioxide. They manage quite well without the special respiratory organs such as lungs, gills, and tracheae that many other animals have. Even though soil flatworms, like potworms and earthworms, each have both male and female reproductive organs, they still mate every so often to exchange sperm and to fertilize each other's eggs. As two flatworms go their own ways after mating, their eggs are left behind in round, leathery cocoons. When newly hatched flatworms leave the cocoons after about a month, they look like miniature versions of their parents and begin feeding on the many protozoa that share the water film with them. After they have grown for a few weeks, young flatworms are ready for larger prey like slugs, potworms, earthworms, springtails, larvae of insects, and even other flatworms.

Since flatworms cannot burrow, they glide through the soil. Squeezing through narrow passages and sliding over leaf litter on their flat bellies, flatworms have been clocked moving along at speeds of 20–50 mm per minute.

The graceful movement of flatworms results from a combination of waving cilia like those that propel the smaller protozoa and slimy mucus like that used by the larger snails and slugs of the soil. For life in the dark and damp environments of the water films that cover soil

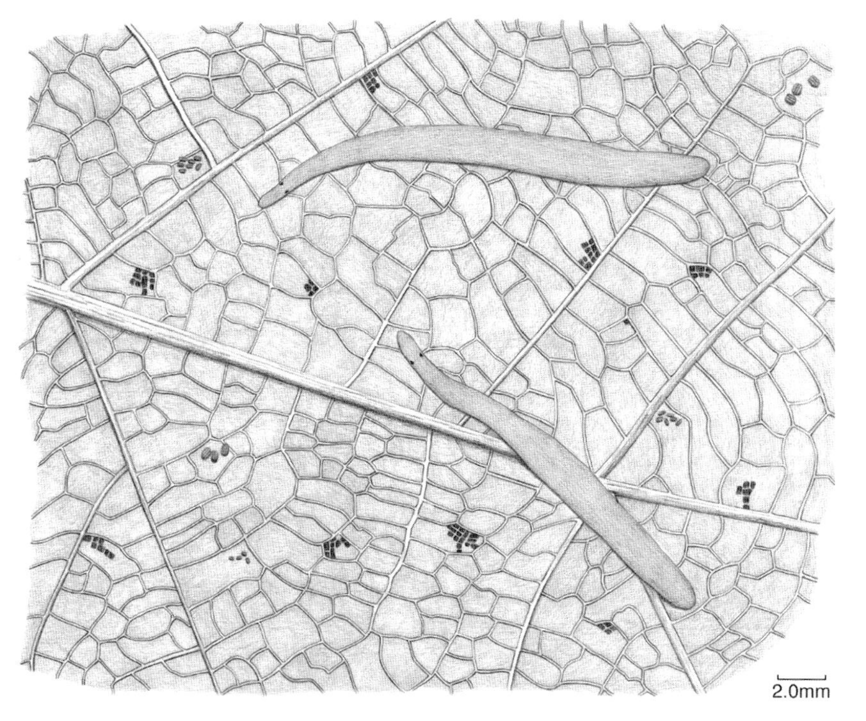

35. Flatworms are some of the simplest multicellular animals of the soil. Their pair of simple eyes helps them avoid light.

and leaf litter, the simple body plan of flatworms has served them very well. Flatworms have colonized soils around the world, wherever they have been inadvertently transported in the soil of potted plants.

2. Roundworms and Potworms

Phylum Nematoda (roundworms)
Place in food web: predators, herbivores, fungivores, parasites
Impact on gardens: allies/adversaries
Size: 0.3–10 mm
Number of species described: 15,000

Phylum Annelida (potworms)
Class Oligochaeta
Order Enchytraeida
Family Enchytraeidae
Place in food web: decomposers, detritivores
Impact on gardens: allies
Size: 5.0–15 mm
Estimated number of species: 600

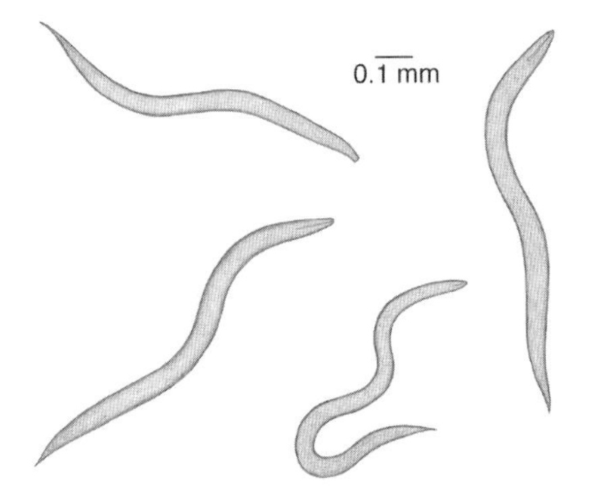

36. Nematodes twist and turn through the innumerable microscopic spaces between soil particles.

Everyone knows earthworms but the other true worms of soils—potworms and roundworms—are smaller, less familiar, and much more numerous. These worms are one and two orders of magnitude smaller than earthworms as well as many orders of magnitude more abundant. A typical earthworm measures about four inches or 100 mm; a potworm or enchytraeid worm measures at most 15 mm; and a roundworm or nematode (*nemato* = thread) averages only one millimeter in length (fig. 36). Grasslands, pastures, and prairies always support the greatest numbers of worms, with cultivated fields supporting only about a tenth of that number. Whereas only 50 thousand to 8 million earthworms can live on an acre of grassland soil, 80,000 million (80 billion) roundworms and 400 million potworms can live in the same area. The total number of individuals representing each type of worm may be very different; but the total weight of each type per acre is comparable, ranging from 100 to several hundred pounds.

Earthworms prefer soils rich in calcium and low in hydrogen ions. Their smaller relatives the potworms, however, are found primarily in soils with lots of decaying matter, and they actually prefer acid soils with high concentrations of hydrogen ions like the soil of pine forests (fig. 37). Nematodes are the least finicky and seem at home just about everywhere in just about every type of soil. In trying to convey an impression of their abundance and their omnipresence, one authority on nematodes, Dr. N. A. Cobb, wrote in the 1914 *Yearbook of the United States Department of Agriculture* that "if all matter in the universe except the nematodes were swept away, our world would still be dimly recognizable . . . we should find its mountains, hills, vales, riv-

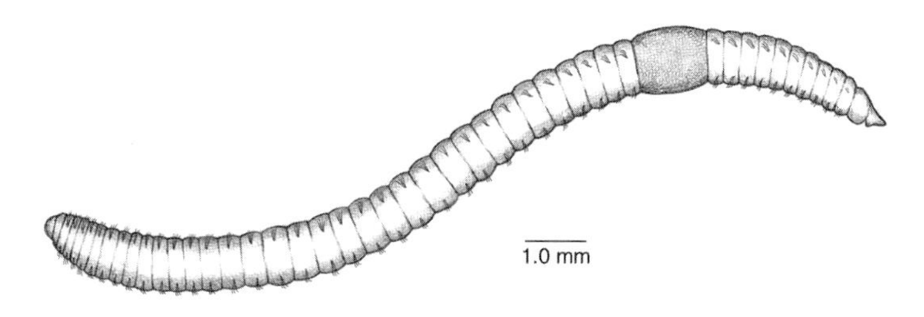

37. A potworm or enchytraeid worm is a diminutive version of an earthworm. It can not burrow like the earthworm but must travel along whatever pores and passageways it can find.

ers, lakes and oceans represented by a film of nematodes." However, our ignorance of the estimated half a million different nematodes that share this earth with us is as vast as the habitat they occupy. Only about 15 thousand, or 3 percent, of these half million nematodes have been described, so many new discoveries and surprises await those who study nematodes.

Earthworms and potworms feast on the plant debris that ends up on the ground, and they quickly transform it to humus. The smaller animals that contribute their share to humus formation, however, often end up as dinner for the smallest of the worms, the nematodes. Nematodes that feed on bacteria and actinomycetes often have an abundance of lips surrounding a narrow mouth that vacuums up food from pores in the soil. Nematodes that are root feeders have mouths with hollow, piercing spears that they ram into roots to tap juices of the plant. The larger nematodes are predators of protozoa, rotifers, tardigrades, as well as smaller nematodes. They are often endowed with as many as six lips at the entrance to their mouths while grinding plates or a few teeth are found just inside their mouths (fig. 38). They must look ferocious and formidable to the humus builders who share their topsoil. Behind each nematode's mouth lies a proportionately large and muscular esophagus; with this feeding pump, the nematode gulps down its meal, whether the meal happens to be root juices or rotifer meat.

There are even some nematodes that attack insects of the soil. A special pouch in the intestines of one of these nematodes carries symbiotic bacteria that act as accomplices in dispatching of insects. When the nematode penetrates the body cavity of an insect host, it releases

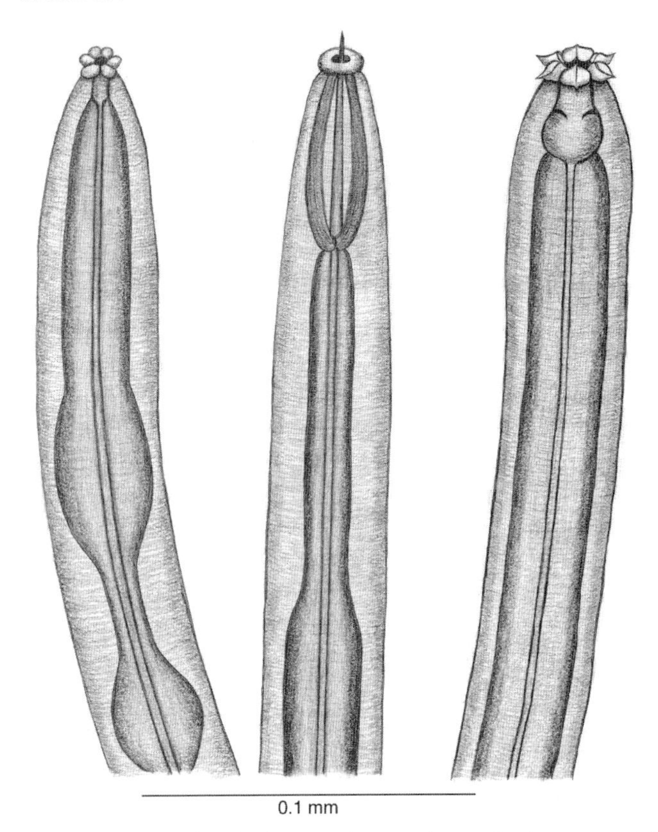

38. A closer look at the heads of nematodes reveals mouths that are specialized for different eating habits. Nematodes that feed on roots and fungi (center) have piercing mouthparts that drain sap. Nematodes that feed on protozoa, rotifers, and other nematodes (right) have large teeth lining their mouths; nematodes that feed on tiny bacteria (left) do not have these imposing teeth.

thousands of these bacteria into the insect's blood. There the bacteria quickly multiply, soon killing the insect but providing a rich source of nutrition for the nematodes. Some farmers are using these nematodes as a natural, nontoxic form of control for such insect pests of soil as beetle larvae and leatherjacket larvae of crane flies.

Nature is always enterprising and opportunistic. As long as nematodes have dwelled in the soil, there have been creatures whose survival has depended on the vast protein-rich resource that nematodes represent. Springtails have been observed grasping nematodes at one end and then sucking them up in a few seconds as though the nematodes were long strands of spaghetti; tardigrades and certain mites probably also eat their share of nematodes. But one of the greatest threats to the well-being of nematodes is certain fungi that have developed not only an appetite for nematodes but also a way to detect their odors and ingenious ways to trap them. There are at least 150 species of fungi that feed entirely on nematodes or that supplement their diets of rotting leaves and wood with tidbits of nematodes. A meal of a nematode or two improves the nutrition of a fungus by adding a good dose of nitrogen to a bland diet of plant debris that is notably high in carbon and low in nitrogen.

Most soil fungi seem deceptively placid as their hyphae meander through the pores of soil, but they can be deadly to nematodes that meander through the same pores. Some of these fungi have sticky spores that latch on to the skin of nematodes. There they germinate and their hyphae begin growing into the nematode. Once the hyphae have filled the nematode and digested its contents, they grow out of the mummified worm and produce more sticky spores that wait for a new nematode to pass by. Another group of fungi produce sticky hyphae rather than sticky spores and catch passing nematodes just as flypaper catches flies. Many of these same fungi produce an alluring odor that attracts nematodes and rotifers, adding to their success as hunters and carnivores.

The most remarkable of the nematode-eating fungi, and also the most common, are snare-forming fungi that have developed the most elaborate traps of all. These traps only form if the fungi are informed in some way that nematodes are in the vicinity. Just the nearby presence of nematodes is sufficient stimulus for the fungi to form the loops; even water that once contained nematodes will trigger formation of traps, showing that the fungi are responding to a chemical made by the nematodes. In some species the inside of each loop is just wide enough for a nematode to slip through part way and then get stuck long enough for the fungus to finish it off by piercing it with its hyphae. Some species of fungi have developed loops that can constrict and strangle a victim. If a nematode should chance to poke its head into one of these loops and touch one or more of the three loop cells, the cells will swell in a flash of a second and grip until the struggling worm wriggles no more (fig. 39).

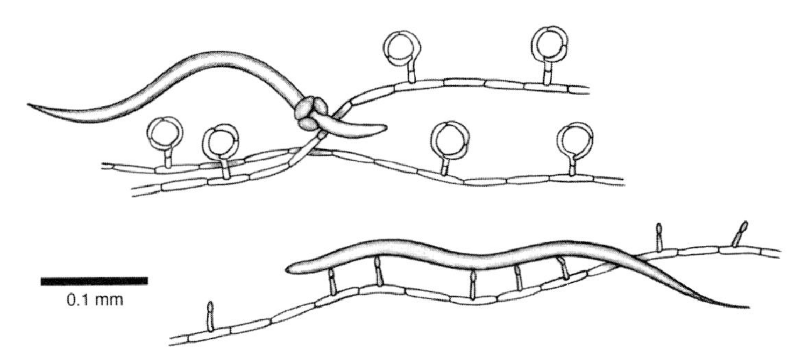

39. Fungi set a number of different traps for nematodes. After a fungus immobilizes a nematode in a noose or sticky trap, it sends its hypha into the nematode's body to digest all the tissues of the nematode except its hard, outer cuticle.

3. Earthworms

Phylum Annelida
Class Oligochaeta
Order Opisthopora
Place in food web: decomposers, detritivores, diggers
Impact on gardens: allies
Size: 10–400 mm
Estimated number of species: 7,260

Every farmer tries to make his land more productive by using fertilizers and the latest equipment for tilling the soil. Although the earthworm works at a much slower pace and on a much smaller scale, every earthworm achieves the same effect as it goes about its daily affairs. As it plows through the soil, often as deep as seven or eight feet (two meters) and occasionally as deep as 10 feet (three meters), an earthworm swallows mineral particles and plant matter, partially pushing its way through the soil and partially eating its way through. Inside the earthworm, mineral particles and plant debris are mixed together and pass out as "castings" or "casts" on the surface of the soil. Most casts are left on the surface of the ground and are especially obvious at entrances to earthworm burrows after a spring rain. As earthworms burrow, they till and mix the soil unlike their smaller relatives the potworms that do not burrow but travel through pores in the soil.

Earthworm manure is always finer, richer, and less acidic than the soil and litter the earthworm initially swallowed. By the time the bits

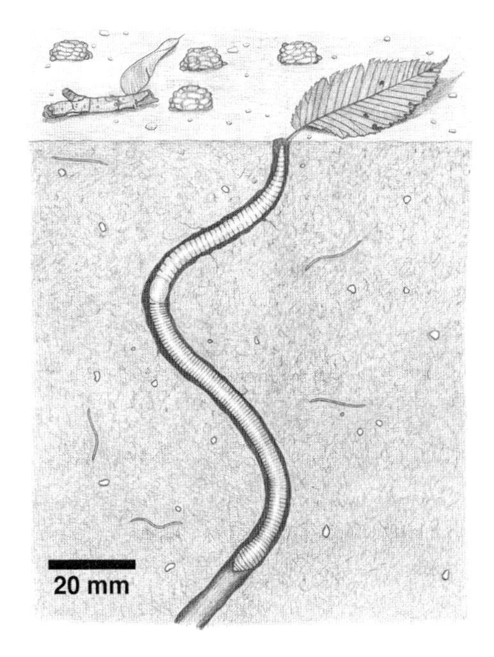

40. Earthworms pull fallen leaves into their burrows and leave their droppings or casts on the soil's surface. In most soils, potworms outnumber earthworms by a ratio of 50:1. Here five potworms (three to the left of the earthworm and two to its right) move through spaces in the soil.

of soil, plant material, and bacteria have passed through an earthworm's gut, many minerals that were previously unavailable to plant roots can now be found in its casts (fig. 40). As it passes through earthworms, soil takes on new properties. Along its long gut each earthworm has three pairs of glands that secrete calcium carbonate ($CaCO_3$), the same compound that farmers add to soil to lower its acidity. Earthworm casts therefore contain around 50 percent more calcium, nitrogen, phosphorus, potassium, and bacteria than the surrounding soil. In this way earthworms contribute to the formation of rich, productive soils that are not too acidic, and earthworms prefer these soils above all others.

In his book *Soil* (1954), G. V. Jacks tells a revealing anecdote about earthworms, moles, and the chalk lines of tennis courts. Dr. Jacks, a soil scientist at Rothamsted Experimental Station in Britain, lived near tennis courts that had been built on acidic soil. The calcium of the chalk ($CaCO_3$) that had been used to make the lines had locally neutralized the acid and created a soil environment that attracted earthworms. Long after the chalk had been washed away by rains, runways of moles still faithfully followed the former chalk lines. The moles had clearly discovered that their favorite food, earthworms, preferred to live along these long, narrow stretches of soil where they were able to avoid the surrounding acidic soil.

Years before earthworms were appreciated as beneficent tillers of the soil, a clergyman living in the small English village of Selborne came to their defense in his book, *The Natural History of Selborne.* When Gilbert White wrote down his observations in 1777, most people had poor opinions of "this small and despicable link in the chain of nature." Gardeners considered them a nuisance for leaving their unsightly casts along garden walks, and farmers believed that earthworms ate seedlings of wheat and corn. How ironic that men who tilled the soil had failed to appreciate what an English clergyman observed so clearly: "Worms seem to be the great promoters of vegetation by boring, perforating, and loosening the soil, and rendering it pervious to rains and the fibres of plants; by drawing straws and stalks of leaves and twigs into it; and most of all, by throwing up such infinite numbers of lumps of earth called worm-casts, which, being their excrement, is a fine manure for grain and grass."

In 1881 Charles Darwin devoted his last book to earthworms and the many observations of them that he had recorded over his lifetime. What he had observed over a long period was that these lowly, legless creatures showed a surprising amount of intelligence—earthworms can actually arrive at conclusions based on their experiences. Earthworms drag fallen leaves into the entrances of their burrows to serve as protection from predators or as part of an evening meal. Earthworms, however, do not just pull any part of a leaf into their burrows. Darwin found that they systematically feel around the edge of a leaf until they find the leaf's tip or its petiole. Some leaves fold and curl more readily at one end or the other while some can be pulled down a worm's burrow equally well from either end. After examining a particular type of leaf, an earthworm can assess which end it should drag into its burrow.

Later in the book Darwin also documented the earth-moving prowess of earthworms by presenting specific examples of Roman ruins and ruins of the prehistoric Britons being buried by many centuries' worth of earthworm casts. Darwin estimated that in places these objects get buried at the rate of a tenth of an inch (2.5 millimeters) each year, with up to 40 tons of casts per acre being added each year to the surface of the soil. Concluding his tribute to earthworms, Darwin wrote, "The plough is one of the most ancient and most valuable of man's inventions; but long before he existed the land was in fact regularly ploughed by earthworms. It may be doubted whether there are many other animals which have played so important a part in the history of the world, as have these lowly organised creatures."

Few people realize that the familiar earthworms of our gardens, yards, and fields are actually immigrants like most Americans, having accompanied humans in their travels from the Old World to the New World. Most undoubtedly arrived as stowaways in potted plants and rootstocks. The smaller, native earthworms have been largely displaced by the larger immigrants, but they still hang on in those remote, uninhabited, and increasingly rarer spots where native vegetation still endures and the soil lies undisturbed. These native species of earthworms work at a measured pace that maintains an abundant supply of leaf litter and humus on the soil's surface.

In contrast, species of earthworms introduced from other parts of the world are now depleting the leaf litter of many forests and eliminating the rich fauna and flora associated with this leaf litter. Within a few weeks these introduced species can recycle and completely remove the layer of leaf litter on the floor of a deciduous forest that normally takes three to five years to decompose. Only recently have we discovered that, depending on particular circumstances, earthworms can contribute either positively or negatively to the fertility of soils as well as to the native plants and animals associated with those soils.

Most of the familiar earthworms of North American gardens and

---

THE EARTHWORM'S DILEMMA

In the dark of night, earthworms issue forth from their burrows to harvest fallen leaves and plant debris aboveground. They avoid the bright light of day and those invisible ultraviolet rays of the sun, to which they are especially sensitive. Even on an overcast day, enough ultraviolet light penetrates the cloud layers to harm a sensitive earthworm after a few hours' exposure. On rainy days, when water floods their burrows, earthworms are faced with drowning if they stay underground or with being burned by the short wavelengths of ultraviolet light if they surface for oxygen. Next time you see earthworms that have emerged from the soil after a rainy night and have perished on sidewalks and in puddles, you will have a better appreciation for the challenges that they face from living in poorly drained burrows.

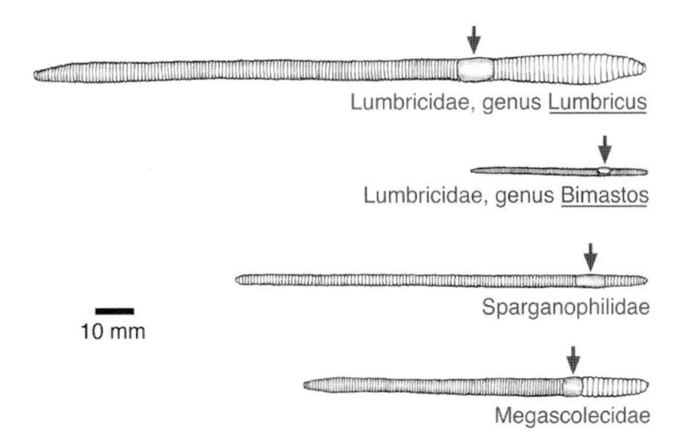

41. A lineup of earthworms from three different families shows three of the important characters that are used for distinguishing one family from another: (1) number of segments between the head and the clitellum (arrows), (2) number of segments spanned by the clitellum, and (3) the shape of the first head segment. The heads are at the right.

farms are introduced members of the family Lumbricidae (*lumbricus* = earthworm). The native earthworms include only a few small members of this family (in the genus *Bimastos*) as well as members of a few other families. Among these native species are mud-dwelling Sparganophilidae (*spargan* = bank; *philo* = love of) found around banks of ponds and streams as well as Megascolecidae (*mega* = large; *scolex* = worm), a family that includes the giant earthworms of Australia, known to reach lengths of three meters.

HOW TO DISTINGUISH ONE EARTHWORM FROM ANOTHER

Over seven thousand species of earthworms are found around the globe, but very few of these species are obviously different in color or shape. All earthworms have segments and bristles on surfaces of these segments, and each adult has a conspicuous clitellum (*clitellum* = saddle), a glandular tissue that produces important substances for mating and egg laying. Assigning an earthworm to a family, genus, or species involves close examination of such features as the shape of the worm's clitellum, the number of segments between the head of the worm and the anterior edge of its clitellum, the particular segments over which the clitellum extends, the shape of a worm's anterior-most segment, and the arrangement of bristles around each segment (fig. 41).

---

4. Land Leeches

Phylum Annelida
Class Hirudinea
Place in food web: predators
Impact on gardens: absent
Size: 8–300 mm
Estimated number of species: 500

Down in the leaf litter and soil live other relatives of the beneficent earthworms. These other segmented worms have developed reputations as infamous bloodsuckers. In southeast Asia and the surrounding islands, horror stories are told of bloodsucking leeches that stealthily emerge from the forest litter to suck the blood of people and other mammals. Most people know about the leeches that live in ponds and streams but not about these leeches of the land.

In Mexico, Guatemala, and southern Europe, the bloodsucking land leeches have not established the sordid reputation of their Asian relatives. These leeches feed on the blood of salamanders rather than the blood of warm-blooded animals like ourselves. While bloodsucking leeches do not live in the soils of the continental United States, these soils, as well as soils in Latin America, Asia, and Hawaii, are host to large land leeches that prey on earthworms and snails. Leeches are most often associated with swamps and ponds, but some of these segmented worms left their aquatic habitats long ago to explore opportunities in the soil. There they discovered a limitless supply of earthworms on which to feast, and they have stayed ever since (fig. 42).

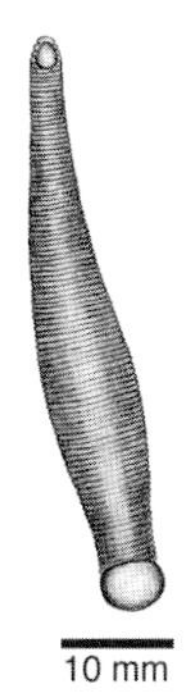

42. Land leeches attach to leaf litter with their larger posterior sucker and wave their anterior sucker about in search of a host.

5. Rotifers

Phylum Rotifera
Place in food web: predators of bacteria and protozoa, scavengers, algal eaters, fungivores
Impact on gardens: allies
Size: 0.1–0.5 mm in length
Estimated number of species: 2,000

After observing these creatures with crowns of cilia waving in circular patterns like the motion of a wheel, early scientists chose the Greek name of "wheel bearers" for them. These animals may be as small as a tenth of a millimeter or as large as half a millimeter. Most rotifers live in ponds, lakes, and streams; but about 5 percent of the estimated 2,000 species have found a home in the soil or on nearby mosses and lichens (fig. 43).

Rotifers usually navigate in the water films that cover the leaf litter, mosses, and soil particles by swimming with their crowns of cilia. As water films shrink on hot, dry days, a rotifer resorts to creeping across surfaces in inchworm fashion: using its head and two toes to contact particles of soil and alternately stretching and buckling the portion of its body between its head and toes (fig. 44). If water films completely disappear, as a last resort rotifers form dehydrated cysts

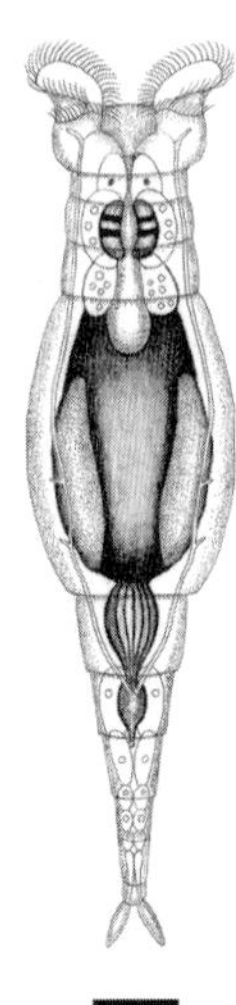

43. *Philodina* is a common rotifer of ponds as well as soils.

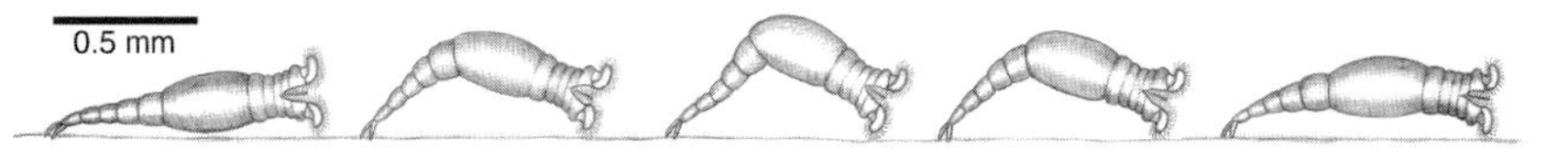

44. Rotifers travel through the soil using the same locomotion as inchworms.

like those of protozoa, tardigrades, and nematodes, which enable them to enter a state of suspended animation until wetter and better times return. During hard times, one particularly common rotifer of the soil called *Philodina* (*philo* = love; *dina* = whirling) transforms to a colorful pink cyst.

For millions of years most species of rotifers have given birth to daughters in a totally female-dominated world. These animals have multiplied and spread without the distractions of courtship and mating. Their principal concerns seem to be surviving droughts, avoiding predators, and finding enough bacteria, algae, protozoa, and plant debris to satisfy their hungers. For rotifers sex is as superfluous as males, and abstinence is a way of life.

6. Snails and Slugs

Phylum Mollusca
Class Gastropoda
Place in food web: decomposers, detritivores, scavengers, some herbivores, and a few predators
Impact on gardens: adversaries/allies

Size: 1.5–120 mm
Estimated number of land snail species: 30,000
Estimated number of land slug species: 500

Like woodlice, which have the distinction of being the only crustaceans to venture from the sea onto the dry land, slugs and snails left their ancestral homes in the sea long before even the ancestors of whales left the land to dwell in the sea, even long before dinosaurs appeared. Snails and slugs are the only mollusks to leave not only their saltwater homes in the oceans but also their freshwater homes in lakes and streams. About 1,000 species are at home in the forests and fields of America north of Mexico. Their needs are simple—a little moisture, some calcium in the soil for producing shells, and a good supply of rotting vegetation. Only a few snails and slugs are predators; they feed on earthworms, potworms, or others of their kind. Each of these soil mollusks is bathed in a slimy, protective mucous

45. Tiny, cone-shaped snails are common scavengers on the forest floor that usually go unnoticed.

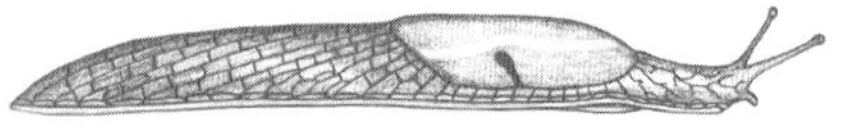

46. Beneath their soft exteriors, slugs carry small, hidden shells.

layer that conserves water and allows each animal to venture across dry stretches of the forest floor without inevitably shriveling and dehydrating (figs. 45–46, plate 15).

Collectively, snails and slugs are known as gastropods (*gastro* = belly; *pod* = feet), creatures whose feet spread out beneath them as they crawl along. Each slug and each snail has one large muscular foot on which it gracefully glides over the swells and swales of the forest floor. As successive waves of contraction pass along the muscles of the foot, the gastropod inches along. Watching the "belly" of a gastropod from below as the animal moves across the clear surface of a jar or terrarium is the best way to appreciate the nimble-footed movements of these animals.

As it consumes plant litter, each animal extends a structure in its mouth that looks and works like a tongue but is covered with many hard, sharp teeth; it is called a radula (*radula* = scraper) (fig. 47). One common snail has 120 rows of 91 teeth each. The teeth of the radula rasp away at decaying leaves, leaving distinctive imprints on a leaf's smooth surface. The material that the radula scrapes from a leaf passes into a digestive tract, where the variety of enzymes involved in breaking down the tougher fibers of leaves and wood is probably greater than that found in the gut of any other animal of the soil.

Gastropods are considered delicacies in many countries, appearing on restaurant menus under their French name, "escargot." Other creatures share our fondness for escargot and have special adaptations for hunting and eating their catch. Certain ground beetles and a couple of families of daddy longlegs have long, powerful mouthparts

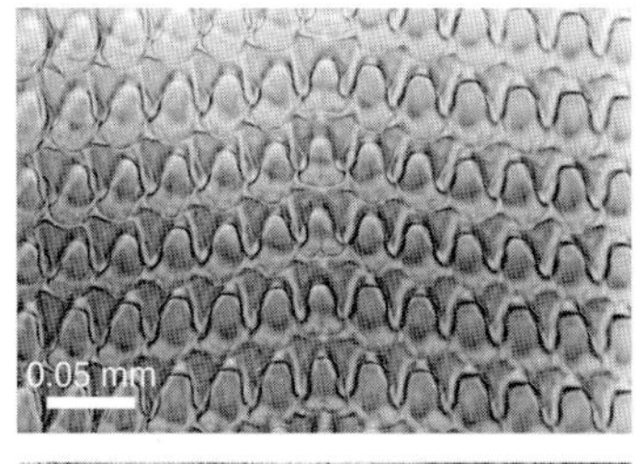

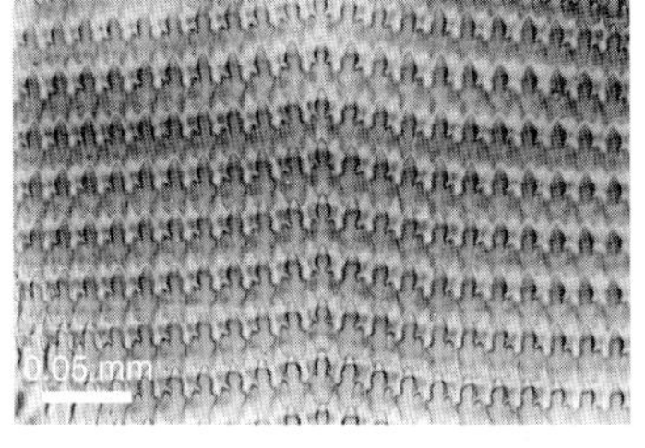

47. The radulae of a snail (top) and a slug (bottom) have precisely arranged teeth that are very effective at rasping and tearing mushrooms and dead leaves.

that enable them to reach deep within the openings of snail shells. The snail-eating ground beetles have narrow, scissorlike mandibles; the jaws of the daddy longlegs are often longer than the rest of their bodies. Glowworms, the larvae of many species of fireflies, can track down snails by following their slime trails. After overtaking a snail, a firefly larva paralyzes it with toxin from its bite. Next the needlelike jaws inject enzymes that begin digesting the snail and enable the larva to suck in juicy escargot. There is even one family of flies whose larvae specialize in being predators of gastropods.

In the woodlands of Britain and Europe, birds regularly smash the shells of snails on rocks in order to get at the meal hidden within. To avoid hungry thrushes, snails have shells with many different color patterns—bands or no bands, yellow or brown—that camouflage them and vary with the snail's habitat. A great range of forms and colors have evolved and are still evolving among the gastropods as snails and slugs continually find new ways to avoid becoming someone's "escargot."

7. Tardigrades

Only in the last two hundred years have we humans even known that such creatures as tardigrades exist. The German naturalist and priest Johannes Goeze published the first account of a tardigrade in 1773, referring to the tiny rotund animal as a "little water bear." Only a few years later, Bonaventura Corti and Lazzaro Spallanzani, two abbots and professors of natural science, contributed their accounts and im-

Phylum Tardigrada	**Place in food web:** predators, fungivores, algal eaters
Class Eutardigrada (naked tardigrades)	**Impact on gardens:** allies
Class Heterotardigrada (armored tardigrades)	**Size:** 0.05–1.2 mm
	Estimated number of species: 400

pressions of tardigrades. Corti referred to them as "little caterpillars," and what caught Spallanzani's attention was the slow, resolute movement of these little animals that walked like turtles. He chose the name *il tardigrado* or "slow stepper." To this day both Goeze's and Spallanzani's names for these animals have stuck.

Water bears are related to arthropods, such as insects and millipedes; but they are unique enough to be placed in their own phylum—Tardigrada. Like arthropods, tardigrades shed an exoskeleton as they grow, and the structure of this cuticle distinguishes the two different types of tardigrades. Heterotardigrades, or armored tardigrades, have thick cuticles that are divided into plates; eutardigrades, or naked tardigrades, have thinner cuticles that may be sculptured but are without any plates. The charm of these little water bears is often enhanced by their appealing colors. Although many tardigrades are colorless or white or brown, some are red, green, orange, yellow, and even pink.

Most tardigrades are about half a millimeter long and dine on just about all the creatures in the soil that are their size or smaller, including protozoa, algae, fungi, rotifers, nematodes, and other tardigrades. There are plenty of larger creatures in the soil that in turn feed on tardigrades—large protozoa, nematodes, carnivorous fungi, mites, spiders, and insect larvae.

At certain times and in certain places, tardigrades can be as abundant as the mites and springtails that seem to be ubiquitous in soil and leaf litter. However, tardigrades tend to prefer certain microenvironments with just the right food, humidity, ventilation, temperature, and other conditions to satisfy their needs. When these conditions are met, as many as 400,000 tardigrades and 1.2 million tardigrade eggs can be found in a square meter of soil. Few eggs are as lovely as tardigrade eggs. They are spherical globes, almost one tenth as large as a full-grown tardigrade, that are ornamented with pores, spines, knobs, ridges, and assorted processes arranged in attractive geometric patterns (figs. 48–49).

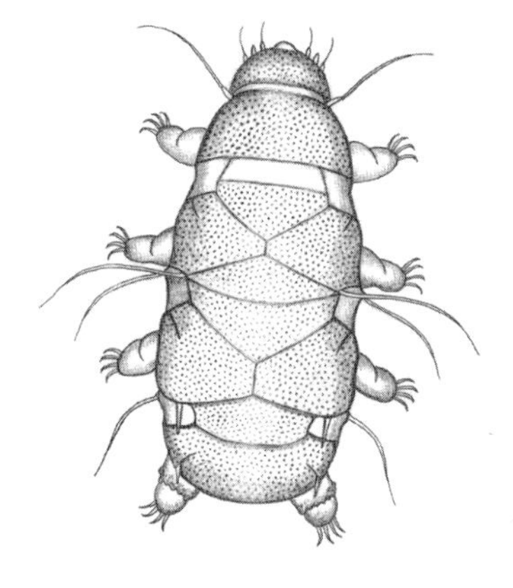

48. Some tardigrades have cuticles with plates and four claws on each leg (bottom). Other tardigrades have thinner cuticles with a few bumps and two double claws on each leg (top).

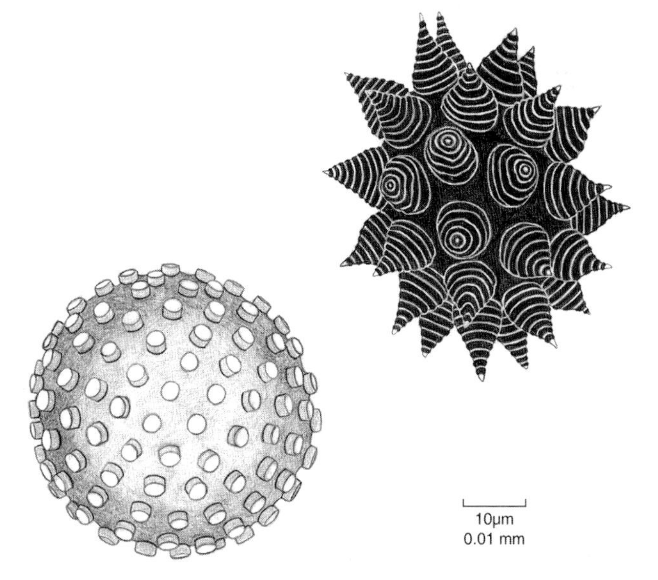

49. The ornate eggs of tardigrades.

Even in their favorite habitats, environmental conditions can often go awry. The temperature can drop too low or rise too high. Drought can lower the humidity to zero. When faced with these dire circumstances, tardigrades have a talent for slowly losing all but about 3 percent of their water and shriveling into forms unrecognizable as tardigrades, yet remarkably still remaining alive. This state of suspended animation has been given the impressive-sounding name of cryptobiosis (*crypto* = hidden; *bios* = life; *-sis* = the act of) and is being actively studied by those who would like to find out if humans can also enter a cryptobiotic state. In their almost completely dehydrated forms, tardigrades can survive high temperatures well above that of boiling water and low temperatures well below −130°C (−200°F). Even after being left on a dried museum specimen of moss for 120 years, a slumbering tardigrade can be awakened from its long sleep.

8. Onychophorans

Phylum Onychophora	**Impact on gardens:** absent
Class Onychophora	**Size:** 15–150 mm
Order Onychophora	**Estimated number of species:** 80
Place in food web: predators	**Lifespan:** 2–4 years

The onychophorans are an ancient group, and their distribution in such widely scattered lands as New Zealand and the islands of the Caribbean suggests that they appeared on earth at a time when the far-flung continents were all part of a single large land mass known as Pangaea. Around 180 million years ago this land mass began to fragment into the continents and islands that we know today, and the onychophorans went along for the ride (fig. 50).

Today onychophorans are barely hanging on in the hot, humid forests of the West Indies, South America, Africa, Australia, New Zealand, and the Malay Archipelago. They hunt for arthropods under rotten logs and in the leaf litter. Just anterior to its mouth, each onychophoran has a pair of glands that explosively expels a sticky material as far as half a meter, entangling and entrapping its arthropod prey. This lifestyle, however, is probably no longer optimal in today's world, as populations of onychophorans steadily continue to dwindle.

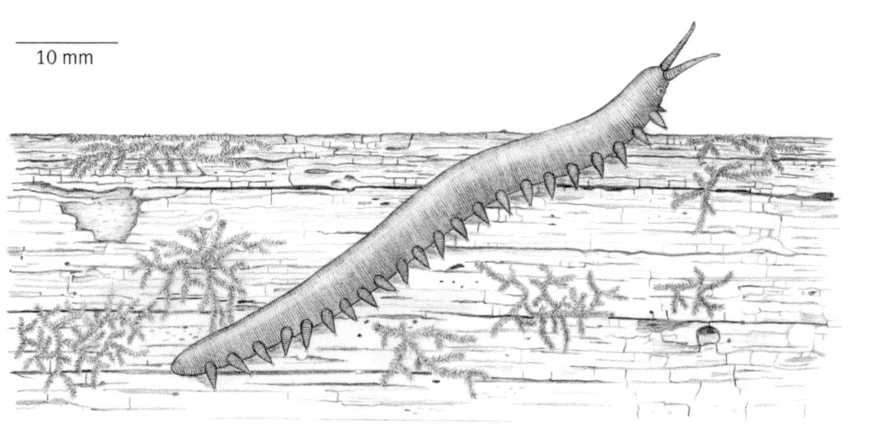

50. Onychophorans are rare animals of tropical forests that share attributes with both earthworms and arthropods.

Although the first biologist to encounter an onychophoran in 1826 described it as a mollusk, some biologists have since then considered them members of the earthworm phylum Annelida, while others have placed them among the spiders, mites, myriapods, insects, and crustaceans of the phylum Arthropoda. Now most biologists consider them such an exceptional and distinctive group of soil organisms that they have placed them in their own phylum, the Onychophora (*onycho* = claw; *phora* = bearers), a name that refers to the curved claws on the ends of their unsegmented legs.

Onychophorans are the best candidates for organisms that represent "a missing link" between the annelids and the arthropods. On one hand, they have the arthropod attributes of dorsal hearts extending from head to tail as well as air tubes known as tracheae that repeatedly branch throughout their bodies. On the other hand, beneath their thin, velvety cuticles are layers of muscles like those of earthworms. Cilia, such as those found on the surfaces of certain protozoa, line the reproductive organs and excretory organs of onychophorans. These tiny hairs, which sweep fluids through their reproductive and excretory ducts, are found in earthworms and other annelids but never in arthropods. For good reasons, onychophorans are considered survivors of an ancient lineage that probably eventually gave rise to both the annelids and the arthropods.

b. Arthropods Other Than Insects

Arthropods are the animals with jointed legs that represent over three-fourths of all animals, and the familiar six-legged insects represent about 85 percent of all arthropods. There are arthropods that differ from insects in the structure of their mouthparts as well as in several other features. A few, such as springtails, diplurans, and proturans, have six legs like insects; some have eight legs like mites, spiders, and scorpions; and the rest have anywhere from ten to 750 legs. Many soil-dwelling insects take to the air as adults. None of the non-insect arthropods has wings.

1. Mites and Springtails

Mites	Springtails
Phylum Arthropoda	Phylum Arthropoda
Class Arachnida	Class Collembola
Subclass Acarina	Order Collembola
Place in food web: decomposers, detritivores, predators, fungivores, algal eaters, herbivores	**Place in food web:** decomposers, detritivores, fungivores, herbivores, algal eaters
Impact on gardens: allies	**Impact on gardens:** allies
Size: 0.1–4 mm	**Size:** 0.1–8 mm
Estimated number of species: 30,000	**Estimated number of species:** 7,500

Arthropods all have exoskeletons and jointed legs, and they are by far the most abundant and diverse animals of soil. They can be as small as a fraction of a millimeter or as large as a hundred millimeters. They can be found in every imaginable habitat the soil has to offer and in just about every habitat found on our planet. In soils of pastures, forests, and cultivated fields, mites and springtails represent the undisputed majority of these arthropods and are the most important producers of humus. Only in arid and semiarid soils do ants outnumber mites and springtails (fig. 51).

To the unaided eye, a mite or a springtail is neither imposing nor prepossessing, but it is amazing how a little magnification can change our views of these creatures. They may be small but they come in an enchanting diversity of forms, colors, and textures (fig. 52).

Wherever mites and springtails are found, mites are usually more abundant than springtails, often several times as abundant. Half of

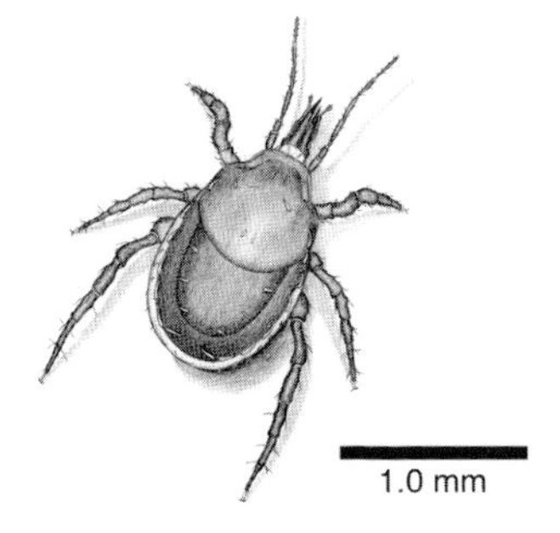

51. This large gamasid mite uses its long jaws or chelicerae to feed on smaller arthropods and their eggs.

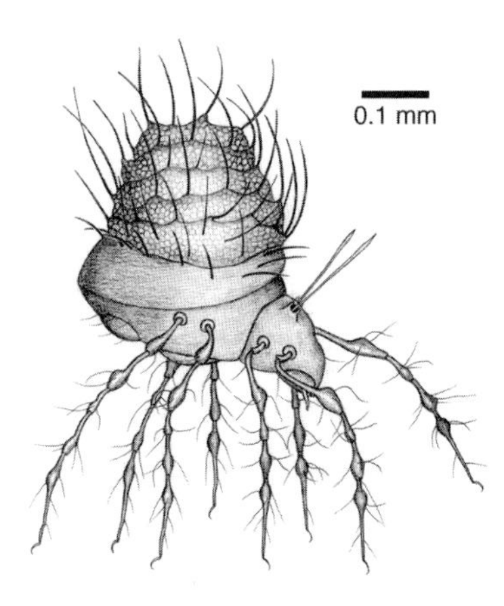

52. This long-legged oribatid mite carries a pile of its shed skins on its back.

the 30,000 species of mites that have been described are dwellers in the soil. Anywhere from 100,000 to 400,000 mites live in a square meter of moist forest soil. Up to 10,000 individuals per square meter are predatory mites, but most mites of the topsoil and litter are the shiny, scavenging oribatid, or beetle, mites. The larger predatory mites have long protruding mouthparts or chelicerae with which they stalk nematodes, potworms, springtails, and other tiny arthropods. The smaller oribatid mites eat fungi and leaf litter with their less conspicuous chelicerae.

The oribatid mites that feed on decaying leaves set the stage for smaller decomposers like bacteria and fungi to free most of the energy and nutrients stored in those leaves. Oribatid mites and fungi, however, do not seem to tolerate each other's company that well. They seem to compete for the same supply of decaying plant litter. Eventually either oribatids predominate or fungi predominate, and bacteria are left to finish the job of decomposing. The mites seem

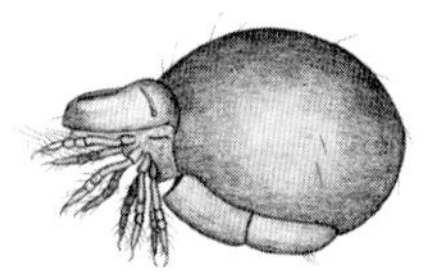

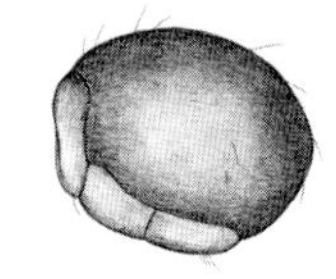

53. No trace of this oribatid mite's legs or mouthparts remain exposed after it closes its shell.

to work best at chewing the fallen leaves and other plant debris into small pieces, making the leaf litter more accessible to smaller decomposers and providing a larger surface area over which microbes can act. Their healthy appetites ensure that each oribatid mite consumes roughly 20 percent of its weight in leaf litter every day. The tiny droppings that they leave behind add to the soil's stockpile of humus.

These darkly pigmented mites of the soil are an exception to the rule that soil-inhabiting arthropods are usually soft, pale, and white. Their dark, hard shells probably serve them well as safeguards from predators and drought. These mites are often referred to as beetle mites or seed mites simply because they look like small, dark, and round beetles or seeds.

Some seed mites are able to close up their shells like clams. With a hinge between the shell at the front end of the body and the shell at the rear of the body, seed mites can tuck in their legs and clamp their front and rear shells shut so that all their appendages are safely covered by their hard shells. The folded mite looks just like a small, round seed (fig. 53).

Those beetle mites that cannot fold their bodies like jackknives resort to other strategies for defense. They have winglike projections on their sides called pteromorphs (*ptero* = wing; *morph* = form). These lateral projections are the closest things to actual wings that any mite has developed. Insects are the only flying invertebrates; but with their pteromorphs, mites came awfully close to having real wings. Pteromorphs have never gotten mites off the ground and into the air, but they have probably saved the lives of a number of mites. Legs tuck neatly beneath the pteromorphs if the mite feels threatened, and there they remain until all seems well again (fig. 54).

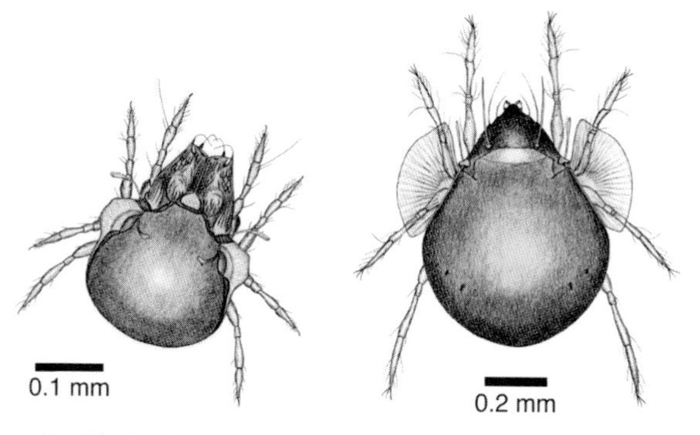

54. Many oribatid mites look as though they have tiny wings. Mites cannot fly. but they can tuck their legs under their "wings" and protect them from the jaws of predators.

Although oribatid mites only lay one or a few eggs at a time, they somehow have succeeded remarkably well in populating the upper layers of the soil as judged from censuses of soil animals in a variety of forests, fields, and pastures. One such census that was taken of the upper 12 inches (30 centimeters) of soil in an English pasture arrived at a figure of 164,000 per square meter for the density of all mites. In the month of May, oribatid mites made up about 60 percent of the pasture's total mite population, while in the month of November oribatid mites represented 45 percent of all mites in the soil samples from the pasture (fig. 55).

Some of the most extraordinary mites can be found far below the soil's surface, where the passageways of the soil are too narrow and too tortuous for most mites. These mites of the deep soil horizons are so long and thin and their legs are so short that they look more like worms than mites. Very little is known about these mites that live in this out-of-the-way, inaccessible habitat. They probably don't encounter many other creatures in the deep soil layers where they roam, so most likely they survive on what little organic debris, such as that left behind by plant roots, finds its way to these depths.

Even with all these mites inhabiting the leaf litter and soil, there still seems to be space for the next most populous group of arthropods, the springtails or the Collembola. Although springtails have diverse shapes, habits, and diets, their tastes in food are not as wide-ranging as those of the mites; and they are more prolific egg layers than the mites. Springtails and mites are both successful, but they

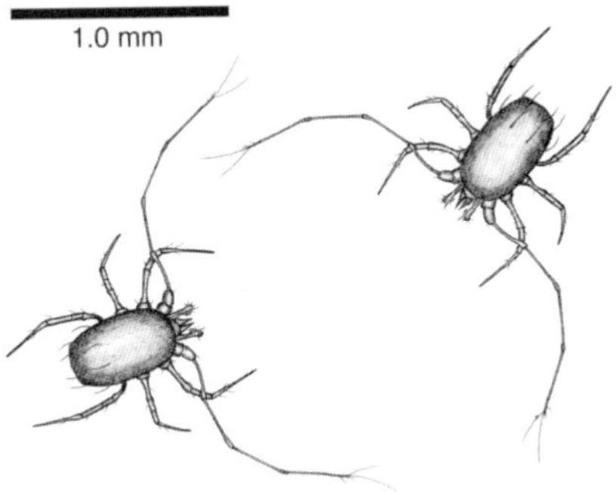

1.0 mm

55. Rather than using their first pair of legs for walking, these small gamasid mites use these exceptionally long legs to explore their environment.

are successful for very different reasons. Each group has developed its own strategy for success. Most springtails are prey and very few are predators. Most make a livelihood as scavengers of decaying plant and animal litter as well as grazers of living mosses, lichens, algae, and fungi (fig. 56).

Encountering springtails frolicking on the snow is probably the best way to get some idea of how extremely abundant these arthropods can be. On winter days, when thick blankets of snow begin melting in the warm sunshine, the water descends into the leaf litter, inundating the countless pores, spaces, and passageways inhabited by springtails. This is when the springtails climb to the top of the snow, and millions of them pepper its surface. Springtails hop about wherever they are, and their antics on the snow have earned them the name "snow fleas," even though they are only very distantly related to fleas. On top of the snow, the winter springtails number around five hundred per square foot; but around tree trunks and in depressions into which they fall, they accumulate by the thousands. The sight of snow fleas on a sunny, winter day gives an idea of the extraordinary number of unseen springtails that normally labor beneath our feet.

Springtails have learned to cope with predators in ways that are different from those of mites. The most obvious difference between springtails and mites is the presence of a springing organ that lies tucked beneath the abdomen of the springtail and can instantaneously launch or "spring" a springtail to safety. With their springs, some animals can jump as much as twenty times their lengths to

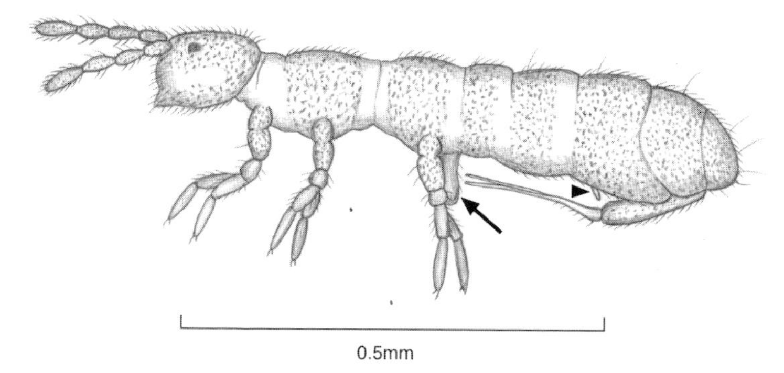

56. In this side view of a springtail, the fork of its tail as well as the catch (*retinaculum,* arrowhead) that holds the tail in place are clearly visible. Its "glue peg" (arrow) is located behind and between its third pair of legs. A springtail might have six legs like insects, but it has half the number of segments in its abdomen that insects have.

avoid the jaws of a predator. Springtails that live on the surface of the soil have ample space in which to leap to safety.

Springtails that live just below the surface in the narrow passageways of the soil do not have quite as much room in which to leap and maneuver. As a result, springtails found in the soil usually have shorter legs, shorter antennae, and shorter springing organs than those that live overhead in the leaf litter. Further down in the soil the pore spaces become even smaller, and springtails appear that have completely lost their springing organs and whose legs and antennae are even shorter. The widths of their bodies are not much smaller than the widths of the pores through which they crawl. Springtails that inhabit deeper layers of the soil have lost their eyes as well as their pigment, but their sense of touch is enhanced by well-developed tactile organs located near the base of each antenna called postantennal organs (fig. 57).

These tiny, blind springtails can navigate through the small pore spaces assisted by their keen sense of touch. They cannot outrun or outleap predators that they encounter, but they can release small droplets of their toxic blood from pores called pseudocelli (*pseudo* = false; *ocelli* = little eyes) that are scattered over their bodies. One taste of this blood soon destroys the appetite of a prospective predator. Ants that attempt to eat one of these pale, helpless-looking springtails are in for an unpleasant surprise. According to one observer of an

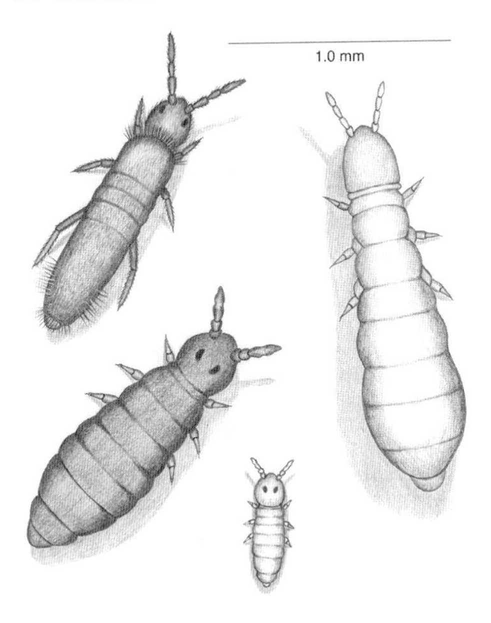

57. The habitat of a springtail can be inferred from the length of its tail, the size of its eyes, and the darkness of its cuticle. The farther a springtail lives from the surface of the soil, the smaller its tail, the smaller its eyes, and the lighter its cuticle. Those springtails that live the deepest in the soil have no tails, no eyes, and no pigment.

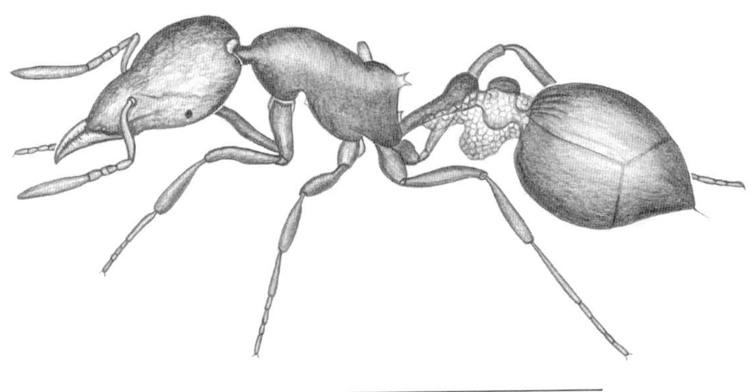

58. This tiny ant of forest soils is a predator of springtails. It lies in wait for a springtail to come within striking range of its large jaws and quickly paralyzes its prey with its stinger. The spongy structure between its thorax and abdomen is believed to produce an odor that springtails find enticing.

ant-springtail encounter, the unsuspecting ants "clean their mouthparts very quickly and most suffer from temporary paralysis in their feelers, mouthparts, and even legs" (fig. 58).

Springtails lack the hard, protective armor that covers the bodies of the oribatid mites and keeps them from desiccating during a

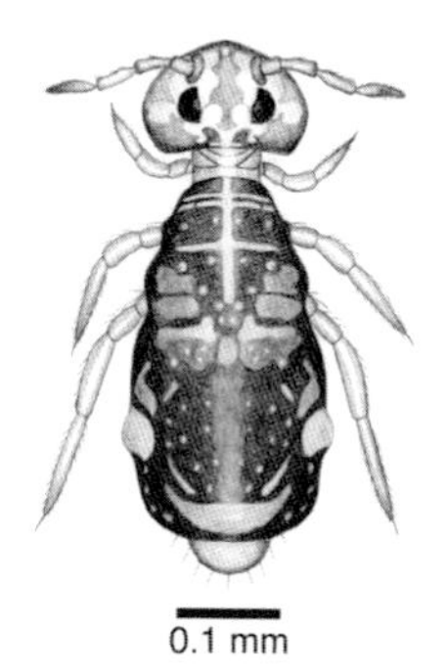

59. This springtail lives in the leaf litter on the forest floor.

drought. Many springtails have thin cuticles covered with scales like those on moths and butterflies; these waterproof structures help the animals retain moisture. A distinctive water-conserving device shared by all springtails, however, is a tube found on each animal's ventral surface that can extend or retract as humidity rises and falls. A springtail walking on the side of a glass container will often check out the moisture in its strange environment by extending this ventral tube. The tube appears to stick to the surface of the glass and explains why the scientific name chosen for the springtails, Collembola, translates to "glue peg" (*coll* = glue; *embol* = peg). Even as leaf litter dries out, springtails are still able to soak up enough moisture with their ventral tubes to survive through the drought.

These various adaptations of springtails to life in the soil are remarkable examples of how animal forms have been molded by their immediate surroundings. Springtails that live only a thin soil layer from each other look and behave very differently. In the soil, microenvironments separated by only a few centimeters can be vastly different (fig. 59).

2. Proturans and Diplurans

Some soil-dwelling arthropods, like insects and mites, have close relatives found in ponds, on trees and flowers, or even as parasites on larger animals. Other groups, such as the diplurans and proturans, have been very conservative about adapting to different habitats. They are pale, eyeless animals that may venture no further from the soil than a rotten log (fig. 60).

Proturans have a distinctive yellow-brown color that makes them easy to spot in a soil or litter sample; however, not until 1907 were they first recognized and described in a publication by the Italian sci-

Phylum Arthropoda

Proturans
Class Protura
Order Protura
Place in food web: fungivores, algal eaters
Impact on gardens: allies
Size: 0.6–1.5 mm
Estimated number of species: 400

Diplurans
Class Diplura
Order Diplura
Place in food web: predators, detritivores
Impact on gardens: allies
Family Campodeidae
Size: 2–7mm
Family Japygidae
Size: 2–50 mm
Estimated number of species: 800

entist Filippo Silvestri. Proturans are smaller than diplurans and rarely grow to more than two millimeters in length, whereas diplurans are about 4 to 40 times larger. Proturans and diplurans resemble insects in that they have six legs, but there most of the resemblance ends. Soil creatures like diplurans and proturans that are without eyes compensate for their loss of sight with enhanced senses of touch and smell. Diplurans have long antennae and equally long sensory appendages called cerci (*cercus* = tail) that project at their rear ends. A variety of sensory bristles are scattered elsewhere over their bodies. What one soon notices about the smaller proturans is that they are missing antennae as well as cerci. A pair of defensive glands at the end of each proturan's abdomen discourages attacks from the rear and makes up for its lack of cerci. Proturans usually walk on only two pairs of legs and use their first pair of legs as antennae. They wave their front legs about just as insects wave their antennae about as they try to locate a particular scent in their environment.

As predators, blind diplurans need a sophisticated sense of touch to enable them to snatch their prey. The smaller and more delicate diplurans called campodeids (*campo* = flexible caterpillar; *-odes* = resemblance to) are known to prey on wriggling midge larvae. The larger and faster diplurans are called japygids (*japyg* = strong wind). The cerci of japygids look like those of earwigs and have taken on the form and function of pincers. Japygids use their pincers to seize smaller campodeids, springtails, and proturans. The long, sinuous bodies of these diplurans give them the ability to whip through the narrow pore spaces of the soil (fig. 61).

60. These pale, blind relatives of insects use their refined sense of touch to navigate through the passageways of the soil. To the right of the 24-legged symphylan is a proturan; below the symphylan is a japygid. To the right of the japygid is a campodeid.

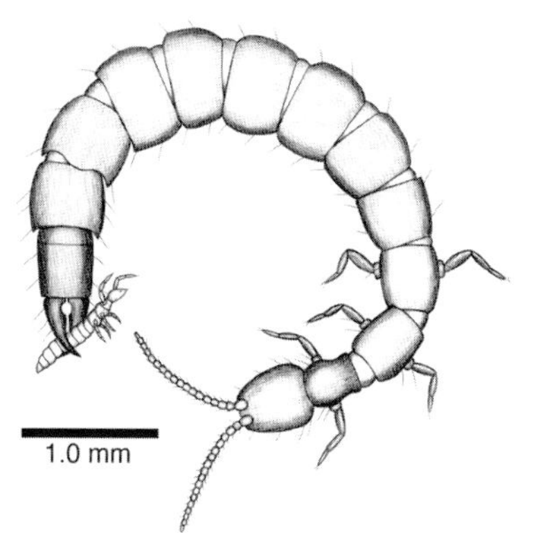

61. A japygid has seized a proturan with its cerci.

3. Myriapods

Phylum Arthropoda

Pauropods
Class Pauropoda
Place in food web: decomposers, detritivores, scavengers, fungivores, algal eaters
Impact on gardens: allies
Size: 0.5–1.5 mm
Estimated number of species: 500

Symphylans
Class Symphyla
Place in food web: decomposers, detritivores, scavengers, fungivores, herbivores
Impact on gardens: adversaries/allies
Size: 1–8 mm
Estimated number of species: 120

Centipedes
Class Chilopoda
Order Geophilomorpha (soil centipedes)
Order Lithobiomorpha (stone centipedes)
Place in food web: predators
Impact on gardens: allies
Size: 3–300 mm
Lifespan: 5–6 years
Estimated number of species: 3,000

Millipedes
Class Diplopoda
Place in food web: decomposers, detritivores, scavengers, fungivores, algal eaters
Impact on gardens: allies
Size: 2–280 mm
Lifespan: up to 11 years
Estimated number of species: 10,000

Nature has been uninhibited about designing legs of soil arthropods. All insects, diplurans, proturans, and springtails have three pairs of legs. All spiders, daddy-longlegs, mites, and pseudoscorpions have four pairs of legs. Woodlice manage nicely with seven pairs of legs. All those arthropods with more than seven pairs of legs are considered myriapods (*myria* = very numerous; *poda* = legs), having anywhere from nine pairs of legs in pauropods (*pauro* = little; *poda* = legs) to 177 pairs of legs in soil centipedes and 375 pairs of legs on a tropical species of millipede. Many arthropods with more than three pairs of legs as adults, however, start out in their youth with only three pairs, adding legs and body segments as they grow and molt. Young larvae of mites, millipedes, and pauropods hatch from their eggs with three pairs of legs and are often mistaken for insect larvae until they molt and develop more legs. The soil centipedes, however, with over a hundred pairs of legs, hatch from their eggs with all, or most, of their legs already formed. With their additional joints and additional legs, the myriapods excel, not necessarily in speed, but in agility and ability to maneuver in tight and twisting corridors.

The smallest, least studied, and most recently discovered myriapods with the fewest legs are the symphylans and the pauropods. These two groups are often the most common myriapods in some soils; yet despite their relative abundance, symphylans were not described until the second half of the eighteenth century, and pauropods were not discovered until the second half of the nineteenth century. Both groups feed on decaying vegetation, fungi, and dead animals; but in other ways, they are very different from one another. Pauropods are only a few millimeters long, have nine pairs of legs, unusual three-pronged antennae, and many long paddle-shaped sensory bristles. Symphylans are about ten millimeters long, have 12 pairs of legs, long, beaded antennae, and a pair of spinnerets on their last segment that spews out silk to foil any predators such as ants that approach from the rear (figs. 62–64).

Biologists have long pondered how the different groups of myriapods are related. Since millipedes are built in many ways like pauropods and centipedes are built a lot like symphylans, biologists have been tempted to think that some time many millions of years ago millipedes descended from ancestors resembling pauropods and centipedes descended from ancestors resembling symphylans. If this is actually the way the myriapod family tree grew, then centipedes and millipedes have gained legs, eyes, and many new colors in the intervening years.

Millipedes feed on plant debris, fungi, and algae, leaving behind numerous droppings that contribute to humus and soil formation. Droppings are good for the soil, and they are good for constructing the nests that most millipedes build for their eggs. Each millipede egg is quite large as eggs go—it can be more than two millimeters in diameter. The mother millipede often envelops each of her eggs in a hollow ball made of soil particles and droppings and then places each of these nests in a pile that she guards very attentively. As newly hatched millipedes leave their nurseries and wander off into the soil, they build new nests of their own for shelter when they are about to molt and shed their thick, protective shells. In addition to their nests and thick shells, millipedes protect themselves with repellent or toxic substances that they release from pairs of defensive glands found on

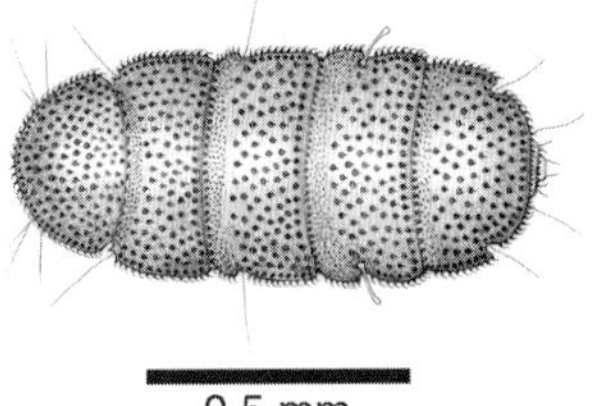

62. Pauropods have been known to science for slightly over a hundred years. Sir John Lubbock, a friend and neighbor of Charles Darwin, first discovered pauropods in his garden south of London.

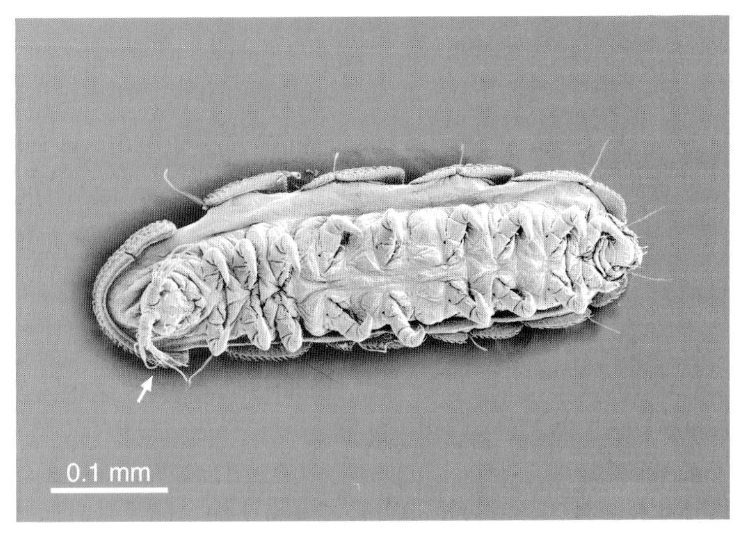

63. This image of a pauropod's ventral surface taken with a scanning electron microscope shows the three-pronged antennae (arrow) and nine pairs of legs on an adult animal.

64. A 24-legged symphylan is often referred to as a "garden centipede" and feeds on roots of garden vegetables.

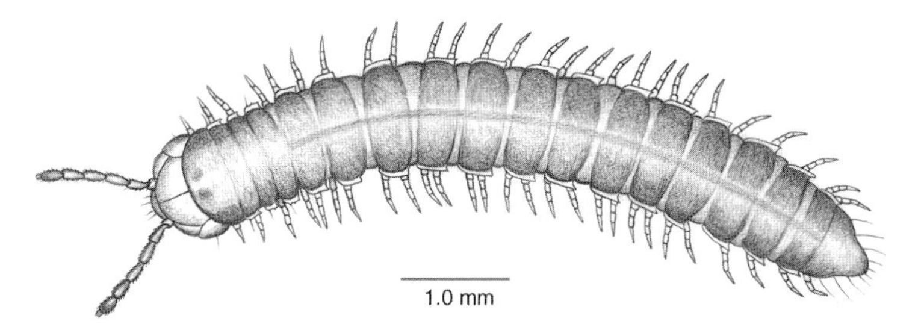

65. Immature millipedes, such as this one, add body rings and legs as they grow and molt.

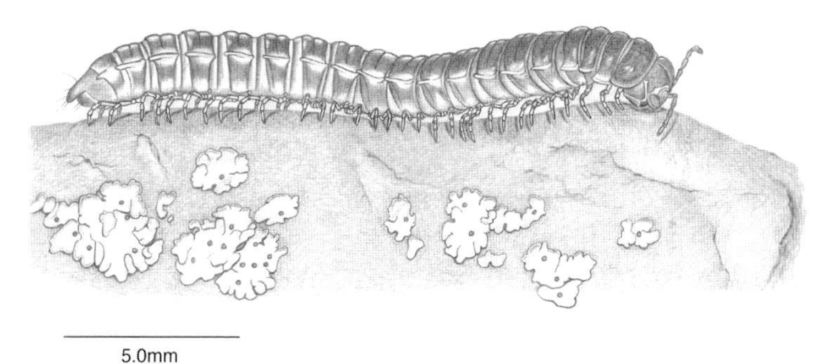

66. Millipedes take a year or more to mature and may live for several years.

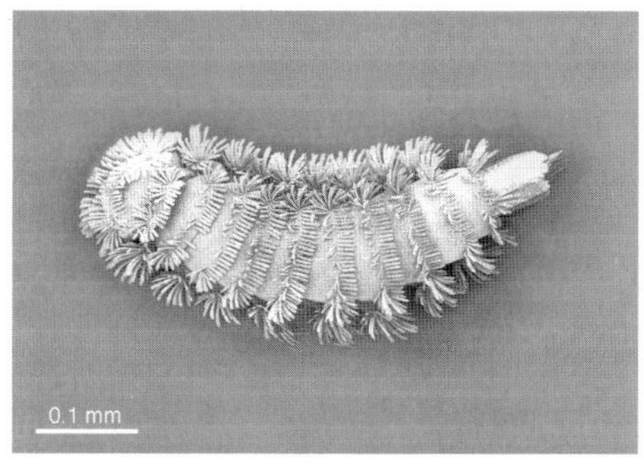

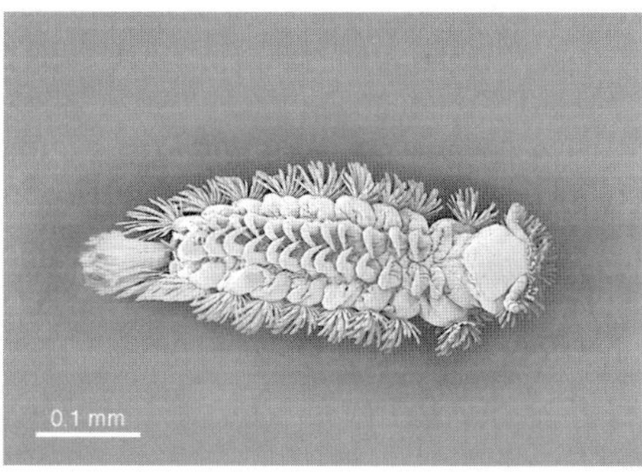

67. Members of one very distinctive subclass of millipedes (Penicillata, *penicillum* = brush) have soft cuticles and are covered dorsally with intricately sculpted scales (top). A tuft or brush of long scales covers its posterior end (right side of top image). On the millipede's ventral surface are its 13 pairs of legs (bottom image).

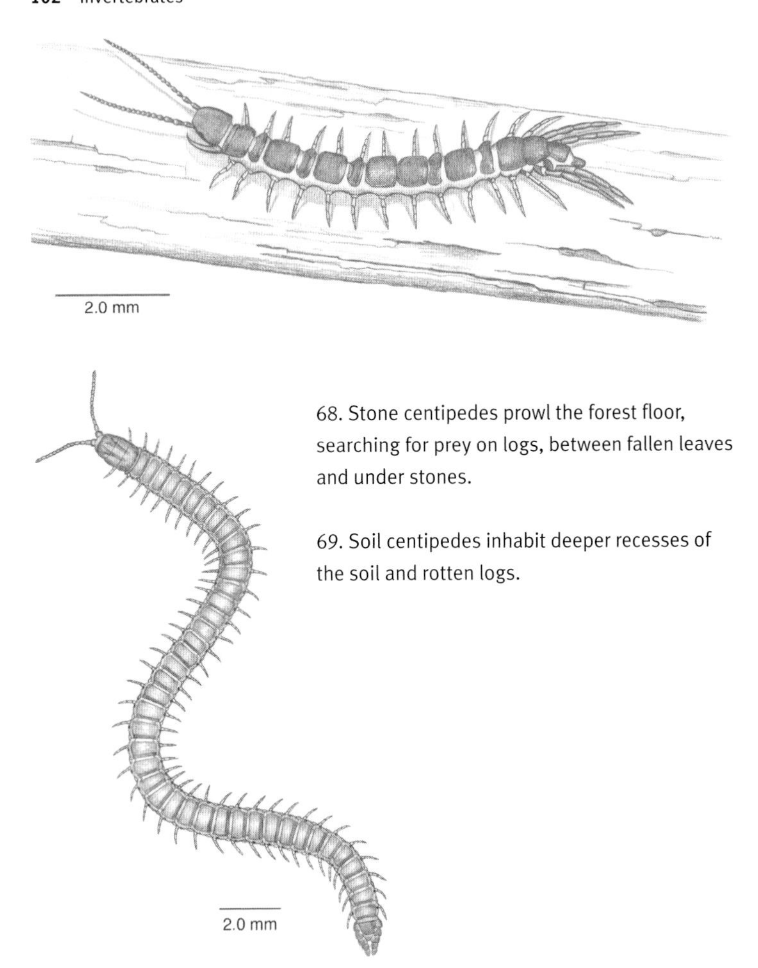

68. Stone centipedes prowl the forest floor, searching for prey on logs, between fallen leaves and under stones.

69. Soil centipedes inhabit deeper recesses of the soil and rotten logs.

the lateral surfaces of most body segments. Millipedes expend a great deal of energy protecting themselves from predators, but they also conserve much energy by not having to search widely for food, such as detritus, that is within easy reach (figs. 65–67, plates 16–17).

Centipedes, on the other hand, must expend considerable energy to stalk and capture prey. A centipede's diet of meat, however, does provide more energy than a millipede's diet of rotten leaves. Enhancing their efficiency as predators, the first pair of legs on centipedes has been specially recruited and remodeled as a pair of poison fangs.

Even though they are predators, centipedes still have to worry about being eaten. Like the symphylans, stone centipedes are prepared for possible attacks from the rear. Symphylans eject silk from spinnerets on their last segment, and stone centipedes have glands on their last pair of legs that expel a similar sticky, stringy substance that quickly gums up the mandibles, antennae, and legs of attackers like ants. Since stone centipedes live on the surface of the soil, beneath rocks, under bark of fallen trees, and in the leaf litter, they are bound to encounter different enemies from those of soil centipedes that live as deep as two feet below the surface. When a soil centipede is assaulted, it quickly draws its long body into a coil and secretes a repulsive fluid from glands on its underside. Soil centipedes are master contortionists that can maneuver their long bodies through the tiny spaces of the soil. They have muscles and flexible joints in places where few other animals do and are able to ferret out potworms and small earthworms from their deep burrows (figs. 68–69).

Mother centipedes guard their eggs and have been observed to lick them. The eggs are susceptible to attack by fungi, and the licking probably coats the eggs with salivary secretions containing natural fungicides. Whatever the ingredients found in these secretions, they are probably more effective and more environmentally friendly than our commercially available fungicides. Soil fungi seem to be a constant threat to eggs in the humid environment of the soil, and maternal care of eggs is fairly common among soil arthropods.

4. Spiders

Phylum Arthropoda
Class Arachnida
Order Araneae
Place in food web: predators, some diggers

Impact on gardens: allies
Size: 3–50 mm
Estimated number of species: 34,000

Spiders are without exception predators, and they have developed a variety of hunting skills. Some, like the swift wolf spiders, simply run down their prey while others spin elaborate webs in the shape of funnels, tubes, or domes that trap their prey. Some of the smallest spiders of the family Hahniidae spin simple sheet webs over tiny depressions of the soil. Most spiders, regardless of their hunting skills

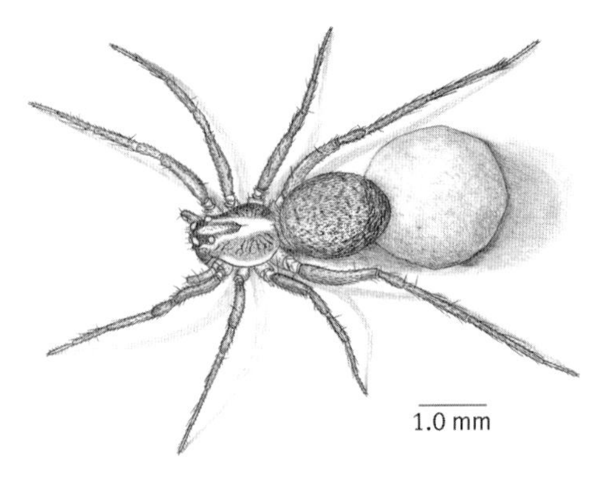

70. A mother wolf spider travels with her egg case over terrain covered with dead leaves, sticks, and stones. Even after her eggs hatch, she carries her spiderlings on her back until they molt for the second time.

or spinning skills, have a retreat in the soil in addition to their web aboveground.

Various estimates of spider densities in forests and fields are in amazingly close agreement. From the leaf litter of a German forest, the number of spiders on a square meter of soil ranges from 50 to 150. On the same area of English pasture, the number of spiders was calculated to be 142. With populations like this, spiders readily keep in check populations of arthropods that grow too large for the available resources.

Spiders are known to shower their young with maternal care. The mother wolf spider carries her egg sac in tow while she hunts, and even helps her newly hatched spiderlings escape from the egg sac by tearing it with her chelicerae, or jaws, when she senses their impending birth (fig. 70). The baby spiders quickly clamber up their mother's legs and ride about on her back until they have exhausted the yolk reserves of their eggs and must hunt for themselves. Baby funnel-web spiders stay in their mother's web for several weeks and coax regurgitated prey from their mother's chelicerae. Young trapdoor spiders stay in their mother's burrow and share her meals for several months after hatching before they venture out to dig their own burrows. Only during this brief debut aboveground does a trapdoor spider leave its burrow. During the rest of its days, it lurks beneath the trapdoor of its burrow. There the spider waits to snatch unwary prey that it can reach with its forelegs without losing a safe grip on the burrow with its hindlegs.

In digging their burrows, spiders use their chelicerae for loosening the soil; then they spin silk around the loose dirt to hoist it from the burrow. Once the burrow of a trapdoor spider is about six inches

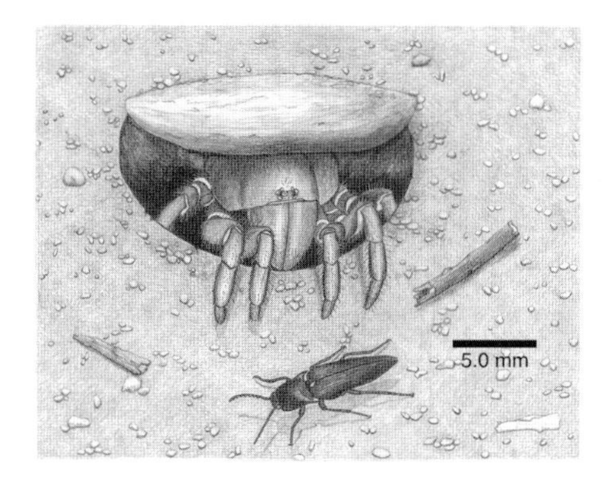

71. A trapdoor spider lies in wait for passersby. In this case a click beetle is about to become a meal for the spider.

(15 centimeters) deep and three-fourths of an inch (2 centimeters) wide, the spider lines all the walls with a smooth layer of silk, fashions a trapdoor with silk that fits snugly at the top, and finally covers it with particles of surrounding soil. The fit of the door and its camouflage are so good that the door is practically invisible when it is closed (fig. 71).

At night wolf spiders can often be spotted on the ground by looking for the red glow of their eight eyes in the beam of a flashlight. This works especially well if you wear the light on a hat. These wolf spiders run down their prey on the surface of the soil; but another group of wolf spiders known as *Geolycosa* (*geo* = earth; *lyco* = wolf) dig shafts as deep as a meter in sandy soils. *Geolycosa,* like trapdoor spiders, hunt from the safety of their burrows. Glowing eyes peering up from tiny holes in the ground probably belong to these burrowing wolf spiders.

5. Daddy Longlegs

Phylum Arthropoda
Class Arachnida
Order Opiliones
Place in food web: decomposers, detritivores, scavengers, herbivores, predators
Impact on gardens: allies
Size: body length 2–12 mm
Estimated number of species: 2,950

Both spiders and daddy longlegs have eight legs, but while practically every spider has the same number of eyes as legs, a daddy longlegs has

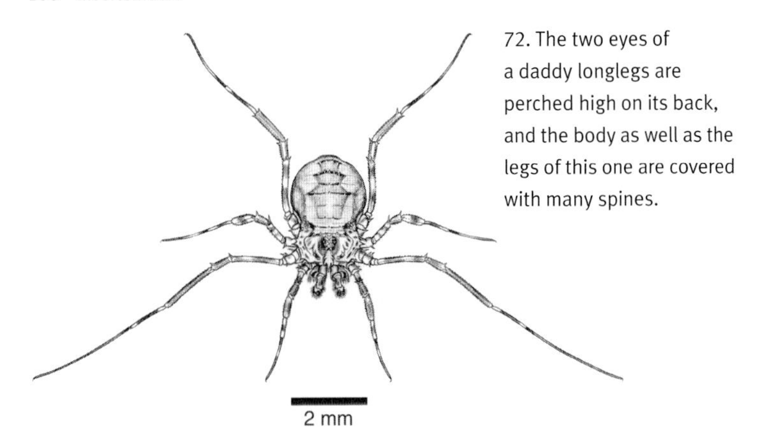

72. The two eyes of a daddy longlegs are perched high on its back, and the body as well as the legs of this one are covered with many spines.

only two eyes, perched almost in the middle of its back. If daddy longlegs were true inhabitants of the underground, they would have a difficult time navigating in the narrow passageways with their eight gangly legs. Instead, they travel on the ground or just above the ground, where they survive on plants, animals, and fungi—either dead or alive. You can often see daddy longlegs waving their legs about, as though testing the air and the objects along their paths. They give the appearance of reaping as they walk, waving their legs to and fro; and this behavior long ago earned them the additional name of harvestmen (fig. 72).

Even though a daddy longlegs that is missing several legs may manage quite well, one that has lost both of its second pair of legs will gingerly pick its way across the ground like a blind person, seemingly unsure of what lies ahead. Unlike insects whose antennae pick up enticing odors of foods and prospective mates, daddy longlegs use their exceptionally long second pair of legs in place of antennae. They are constantly grooming and preening these long, sensitive legs by pulling them between their jaws to remove any debris that could possibly diminish their sensitivity.

Although daddy longlegs are meticulous about their leg grooming, they often end up with small hitchhikers that latch on to these legs (fig. 73). By climbing aboard a passing leg, mites and pseudoscorpions set off for new territories and colonize new habitats. Their own short little legs could never carry them as far as those legs of a passing daddy longlegs, and daddy longlegs do not seem to mind accommodating their small passengers.

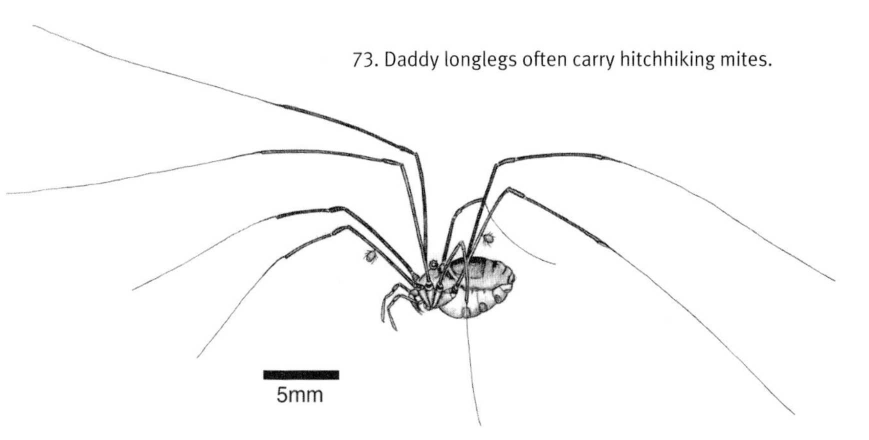

73. Daddy longlegs often carry hitchhiking mites.

The trip of such a hitchhiker is almost assured of being a safe one, for each daddy longlegs is well protected from larger animals that are always looking for a tasty meal. A daddy longlegs can certainly not offer a tasty morsel since it secretes a foul-smelling fluid from two glands, one located below its only pair of eyes and the other behind its first pair of legs. Spiders have been observed attacking daddy longlegs but then quickly backing off to wipe the disgusting taste from their jaws.

Several daddy longlegs that are confined to the same chamber must share their potent odor, and even they are not insensitive to its influence. The substance secreted by the two glands of each daddy longlegs has an anaesthetizing effect on the confined animals. They all remain in a stupor until their odors are diluted by some fresh air, and the daddy longlegs can eventually amble off.

6. Pseudoscorpions

Phylum Arthropoda
Class Arachnida
Order Pseudoscorpiones
Place in food web: predators
Impact on gardens: allies
Size: 2–7 mm
Estimated number of species: 2,000

The best way to become acquainted with pseudoscorpions is to look for them as they fall from leaf litter or seasoned manure that has been

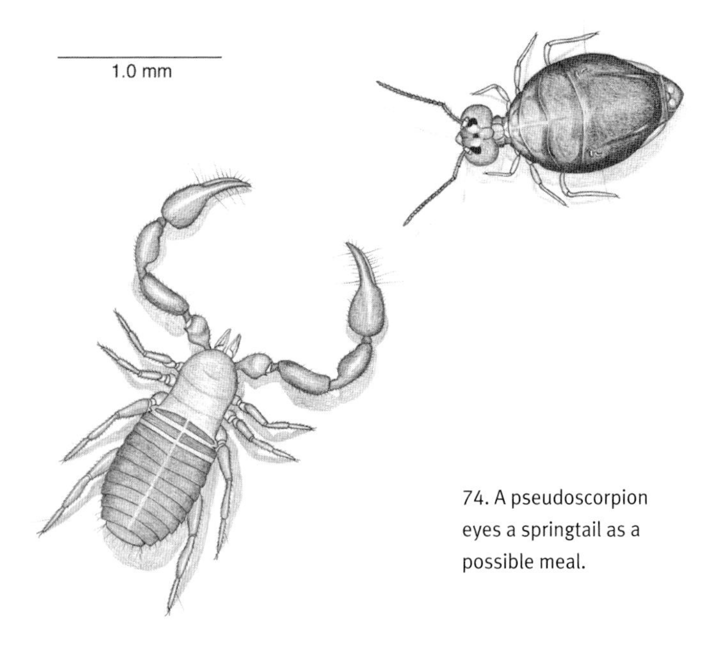

74. A pseudoscorpion eyes a springtail as a possible meal.

placed in a Berlese funnel. As they drop into a container below the funnel, they will draw in their eight legs and remain motionless after the shock of being suddenly transported to this new, strange environment. Within a few seconds, however, they muster the courage to begin exploring the alien world around them. They stretch their legs and hold their large pincers in front of them like the antennae of an insect as they begin to walk slowly across the bottom of the container with the deportment of creatures on some very important mission.

There happen to be at least 2,000 species of these stingless little scorpions that most people never see or even know exist. I usually find only a few when looking through leaf litter, but once I found thousands in a pile of straw and manure. Pseudoscorpions are attracted to habitats where springtails, their favorite prey, are most abundant (fig. 74). Pseudoscorpions usually ignore the more numerous but better-armored oribatid mites of the soil. In the dark passageways of leaf litter and compost, a pseudoscorpion stretches forth its pincers like a person groping about in the dark. Long, touch-sensitive hairs on the pincers signal the presence of a springtail. Instantly it is recognized as a meal and seized with pincers and mouthparts known as chelicerae.

The elaborate mouthparts of this simple creature are not only involved in eating; the chelicerae also function as the source of the silk with which pseudoscorpions mix small pieces of sand and plant debris to build igloo-shaped cocoons. In these silken chambers they spend the unpleasant days of winter and their vulnerable days of molting. By spinning its silk from its mouthparts, a pseudoscorpion's artistry resembles that of its distant relatives, the caterpillars, more than it does that of its close relatives, the spiders. A caterpillar spins silk from its mouthparts while a spider spins silk from the tip of its abdomen.

Some silk-spinning creatures can use their silk as a means of transportation. An inchworm caterpillar can drop from one branch to another as it draws out a silk strand from the end of its spinneret, and a spiderling can sail off into the air when a gust of wind catches the strands of the diminutive silk balloon that fans out from its spinnerets. Pseudoscorpions, however, have not discovered a way to use their silk for transportation. They use little or no energy for transportation; instead they rely on others to supply the energy for them. What they do is hop aboard the long legs of some large passing insect or daddy longlegs and rely on arthropods with longer strides than their own to carry them to destinations they could never reach on their short, tiny legs. Thanks to this form of public transportation provided by larger arthropods, hardly a patch of leaf litter in the forest is without its population of pseudoscorpions.

7. True Scorpions, Windscorpions, Whipscorpions, and Schizomids

Of all the animals of the soil, only the scorpion has the distinction of a having a constellation of the zodiac named for it. The path that our sun appears to follow through the stars passes through a series of twelve constellations that together are known as the zodiac. Scorpius, on the southern horizon in the Northern Hemisphere, happens to be one of those few constellations that is really shaped like the creature it is supposed to depict. Woe to those born under the sign of Scorpius, a sign associated with war, plagues, and storms; and woe to the reputation of the poor scorpion, which really prefers to hide rather than fight and stings only in self-defense.

Several arachnids go by the name of scorpion, but they are considered different enough to be placed in orders of their own. There are whipscorpions and windscorpions, true scorpions and pseudoscorpi-

Phylum Arthropoda
Class Arachnida
Place in food web: predators, some diggers
Impact on gardens: allies/absent

True Scorpions
Order Scorpiones
Size: 40–170 mm
Estimated number of species: 700

Windscorpions
Order Solifugae
Size: 10–50 mm
Estimated number of species: 800

Whipscorpions
Order Uropygi
Size: 25–80 mm
Estimated number of species: 110

Tailless Whipscorpions
Order Amblypygi
Size: 8–45 mm
Estimated number of species: 136

Schizomids
Order Schizomida
Size: 5–7mm
Estimated number of species: 35

ons, and there are tiny relatives of scorpions called schizomids whose place in the family tree of arachnids is far from certain. Unlike pseudoscorpions, which live in most climates as long as they can find shelter under bark and stones, in the soil, and in leaf litter, other scorpions are found in warm, often hot, climates. Whipscorpions and schizomids prefer warm and humid habitats. Some true scorpions also live in warm, humid places; but others are found in the hot, dry deserts that are home to windscorpions (fig. 75, plates 18–21).

Most scorpions can burrow, and they sometimes dig several feet beneath the soil's surface to escape the heat and light of the sun. Many also dwell under stones or logs, in crevices of the earth, or even in burrows of larger animals.

With their poor vision, scorpions depend largely on their sense of touch to maneuver in their dark surroundings. Arachnids might not have antennae, but they have resorted to using a number of different appendages in lieu of antennae.

Arachnids have pedipalps (*pedi* = foot; *palp* = feel) at their head end that are just one of several pairs of appendages that they can use as feelers. True scorpions, pseudoscorpions, and whipscorpions use these pedipalps as strong pincers as well as sensitive feelers; they rely on the long, sensitive hairs of their pedipalps as their main organs of touch. The pedipalps of schizomids are covered with spines and look more like the raptorial legs of a praying mantis than the pincers of

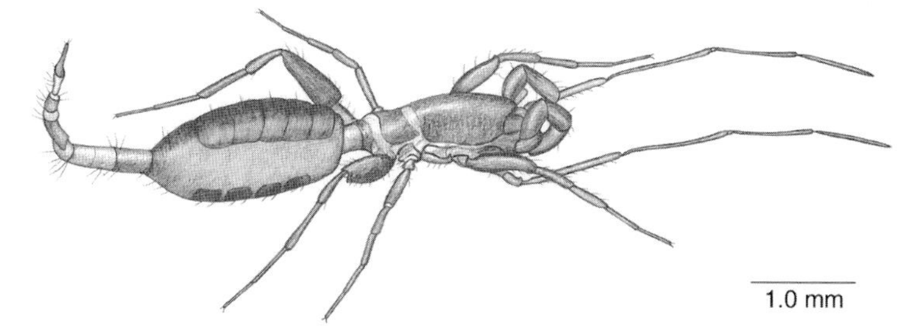

75. Schizomids are tiny, blind arachnids that live under rocks and logs and in leaf litter.

other scorpions. Windscorpions, however, use their enormous jaws or chelicerae as pincers and their pedipalps strictly as feelers. Many arachnids also use some of their legs as feelers. True scorpions and pseudoscorpions use all eight of their legs for walking, but whipscorpions, windscorpions, and schizomids wave their first pair of legs about—just as daddy longlegs wave their second pair of legs—investigating their environments as they move through the leaf litter. Arachnids have perfected their skills of navigation without any antennae and often without any eyes.

All scorpions watch over their young until the little scorpions reach their first molt. Many dig deep burrows for egg laying. Whipscorpions and windscorpions have nurseries that can be 40 centimeters underground. True scorpions give birth to live young that soon find their way to their mother's back, where they ride about during the next two weeks of their lives. Each mother pseudoscorpion and each mother whipscorpion carries about two dozen eggs in a sac under her abdomen. The young hop aboard the backs of their mothers as soon as the eggs hatch. Although mother windscorpions carry neither an egg sac on their bellies nor newly hatched young on their backs, many of them lay as many 100 to 200 eggs in a deep burrow and closely guard the young windscorpions until they shed their first exoskeletons. A mother schizomid builds a small chamber in the earth, where she lies with her few eggs until they hatch. The time invested by these arachnids in maternal responsibilities probably saves many young from jaws of predators and filaments of fungi.

8. Microwhipscorpions

Phylum Arthropoda
Class Arachnida
Order Palpigradi
Place in food web: predators
Impact on gardens: absent
Size: 1–3 mm in length
Estimated number of species: 80

Pale, blind, and covered with many sensory bristles, microwhipscorpions, or palpigrades, are consummate examples of creatures that spend their entire lives underground. Because it uses its two pedipalps for walking, each palpigrade (*palpi* = feeler; *gradi* = walk) appears to have ten legs. Its long, conspicuous tail or whip is a third to half of its entire length, and it is coated with numerous sensory hairs like the bristles of a bottlebrush. Each mother palpigrade invests her energy in only a few eggs at a time, and, like the larger mother scorpions, may diligently guard her small brood. Relatively little, however, is presently known about the lives of these secretive arachnids, which are found in moist soils around the world, except at cold latitudes and high altitudes (fig. 76).

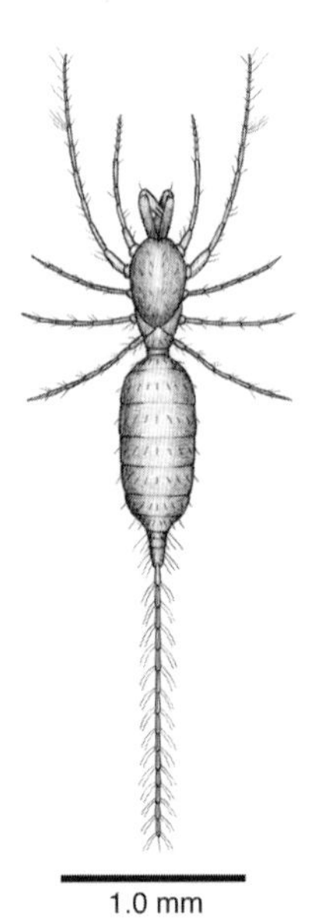

76. Some of the tiniest and least studied arachnids are the microwhipscorpions.

9. Ricinuleids

Phylum Arthropoda
Class Arachnida
Order Ricinulei
Place in food web: predators
Impact on gardens: absent
Size: 5–10 mm
Estimated number of species: 55

Ricinuleids (*ricinus* = a kind of tick) are uncommon arachnids that have the common name of hooded tickspiders. They look like ticks and have jaws like spiders, but they have a hinged, cuticular hood covering their jaws that is unique to their order. The hood can be raised and lowered during feeding. A ricinuleid grabs its prey with the relatively small pincers of its pedipalps, lifts its hood, and then passes the prey to its jaws. The hood is lowered as the ricinuleid chews away.

What little has been observed about the specific courtship and reproduction behavior of ricinuleids has not been observed for any other arthropods. During courtship males use their modified third pair of legs to transfer sperm to the female. The fertilized eggs are then transferred to the shelter of the female's hood. A mother ricinu-

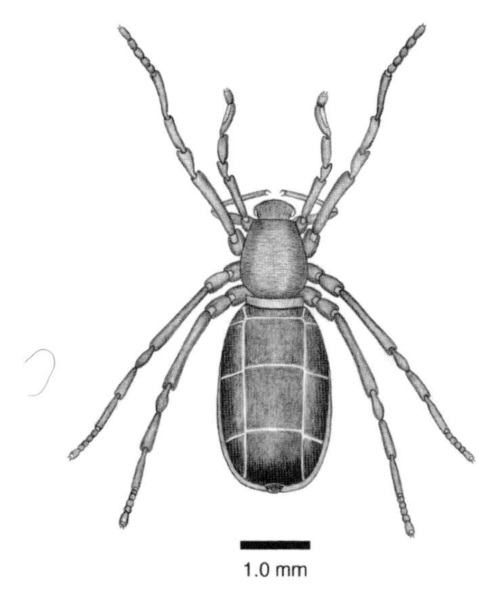

77. Ricinuleids, like ticks and mites, are the only arachnids that hatch from their eggs with six legs rather than eight.

leid carries her few fertilized eggs until the young hatch with six of the eight legs they will have as adults.

These predators are found in leaf litter of warm, humid environments. In the Old World they are known only from Africa; in the New World they are found as far south as Brazil and as far north as Texas. The natural history of these tropical, and often reclusive, arachnids is still largely unexplored (fig. 77).

10. Woodlice

Phylum Arthropoda
Class Crustacea
Order Isopoda
Suborder Oniscoidea
Place in food web: decomposers, detritivores, scavengers, coprophages, algal eaters
Impact on gardens: allies
Size: 1–30 mm
Lifespan: 2–3 years
Estimated number of species: 1,000

While insects are the undisputed arthropod masters of the land, crustaceans are the dominant arthropods of the oceans. Once upon a time all crustaceans lived in the sea, but a few managed to colonize freshwater habitats and eventually the land. Although these pioneers have forsaken their homes in the water, they have found ways to conserve and carry water with them (fig. 78).

Water-conducting channels run around the body of each woodlouse and beneath the surface of each body segment. Urine from the woodlouse's equivalent of kidneys is recycled through these channels and additional water from the environment is taken up as needed by capillary action, just as water is automatically drawn up the channel formed by a straw that is placed in a glass of water. Ammonia from the urine quickly evaporates, and oxygen is readily absorbed at the posterior end of each animal by its pleopods (*pleo* = full; *pod* = leg), remnants of its aquatic heritage. Beneath the outer surface of the pleopods, woodlice have retained gills that are bathed by the recycled water that constantly circulates around their bodies (plate 22).

As an additional adaptation to life in dry soils, some woodlice also have oxygen-absorbing surfaces on their pleopods that look and act like tiny lungs. These "lungs" are not bathed with oxygen-

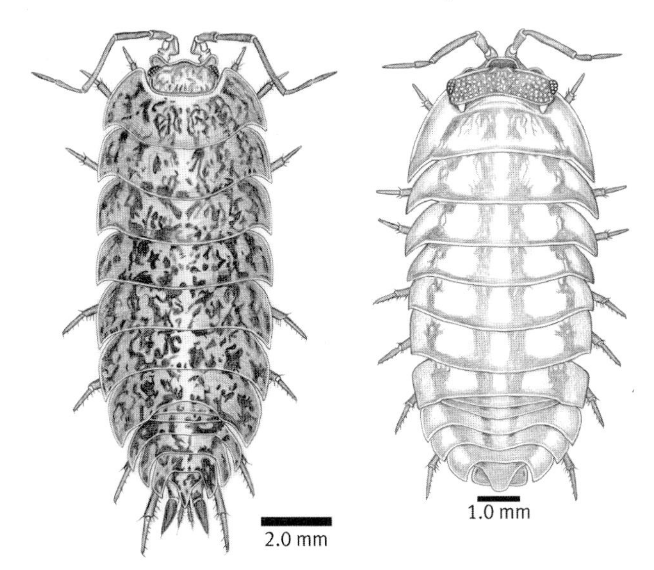

78. Woodlice have been named after both armadillos and pigs. Woodlice that roll into balls belong to the family Armadillidiidae (see fig. 12). Woodlice known as sowbugs, probably because they root around in the soil like pigs, belong to the family Porcellionidae (*porcelli* = a little pig).

enriched water but take oxygen directly from the air just like our own lungs.

In the soil community, woodlice begin the process of converting freshly fallen leaves to humus. They quickly skeletonize and fragment these leaves, setting the stage for beetle mites, certain springtails, potworms, and bacteria to continue the process of humus formation. Woodlice clearly prefer the flavor of new-fallen leaves to leaves that have spent the winter on the ground.

Not only are woodlice primary decomposers of the leaf litter, but they are also secondary decomposers of each other's dung. An interesting physiological phenomenon underlies this behavior. The oxygen-absorbing pigment in the blood of woodlice contains copper, an element reasonably common in their ancestral seas but relatively uncommon in most land environments; it is lost from a woodlouse's body every time it defecates. Just as they recycle urine in their water-conducting channels, woodlice recycle their droppings and the precious copper that they contain.

11. Crayfish

Phylum Arthropoda
Class Crustacea
Order Decapoda
Place in food web: diggers, scavengers, decomposers, detritivores, predators
Impact on gardens: absent
Size: 80–100 mm
Estimated number of species: 150

Crayfish can be abundant in low-lying forests and fields, where they dig deep holes in the moist soil. Although crayfish are rarely as numerous as earthworms, one crayfish can make a significant contribution to the turnover of soil. Crayfish never stray too far from water, and their vertical tunnels always descend as deep as the underlying layer of soil that is saturated with water. Since this layer, known as the water table, may lie as deep as four to five meters below the surface, excavating a shaft to the water table can bring many other soil layers to the surface, each with a different texture and different minerals. During the digging of its tunnel, a crayfish pushes soil to the surface and piles it in an orderly fashion around the hole. Eventually the pile can grow to a chimney that is 20 centimeters tall and that may weigh as much as two kilograms. On an acre of ground, several hundred chimneys may dot the landscape at one time. The weight of all those chimneys adds up to over a ton of soil that is carried to the surface of a single acre by crayfish alone (figs. 79–80, plate 23).

c. Insects: The Most Abundant Arthropods

Insects have colonized the earth, the air, and the waters. They feed aboveground as well as belowground and are the only invertebrates that can fly. Most of the soil animals discussed so far are considered euedaphic (*eu* = true; *edaphos* = soil), or true, lifelong residents of the soil, not just transient, seasonal residents. Many insects, however, spend only a portion of their lives in the soil. Some, like grasshoppers, begin life as eggs in the ground; but once they hatch, grasshopper nymphs take up residence on leaves and stalks. Others, such as cicadas, hatch from eggs laid in trees and grasses, but the newly hatched nymphs quickly fall to the earth and remain underground for as long as 17 years. Those that dwell in decaying logs, leaves, dung, or the re-

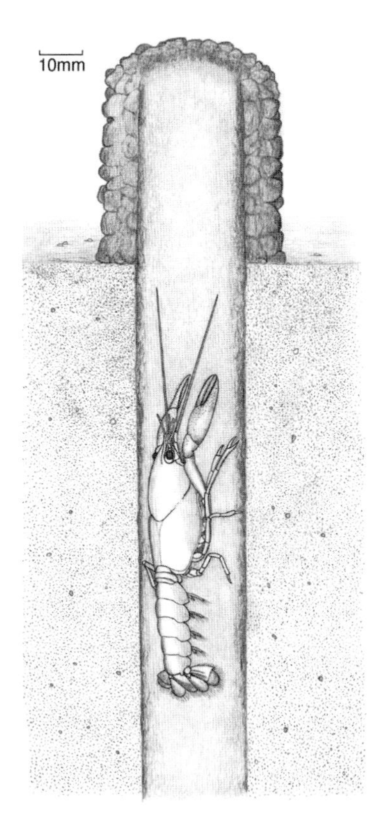

79. Beneath crayfish chimneys lie tunnel systems that are often a meter deep. Poorly drained soils avoided by earthworms and ants is often well mixed by the digging of crayfish.

80. The mound of soil excavated by a crayfish adds new nutrients to the topsoil of a field. (Michael Jeffords)

mains of animals might not actually live underground but they are preparing their habitats to become part of the soil.

1. Jumping Bristletails and Silverfish

One of the first features a person notices about jumping bristletails is their large, bulging eyes. These insects are clearly animals of the soil surface and not dwellers in the dark underground. As grazers in the leaf litter, they feed on algae, fungi, and lichens found among decomposing leaves, where they very effectively evade the many preda-

Phylum Arthropoda
Class Insecta

Order Microcoryphia (jumping bristletails)
Size: 10–12 mm in length
Estimated number of species: 350

Order Thysanura (silverfish)
Size: 10–12 mm in length
Estimated number of species: 400

Place in food web: decomposers, detritivores, scavengers, fungivores, algal eaters
Impact on gardens: absent

tors of the soil surface. Jumping bristletails jump several times their lengths by quickly and firmly slapping the ends of their abdomens against the ground and springing to safety (fig. 81).

Silverfish are similar in appearance to bristletails; however, their bodies are flatter, wider and their eyes are smaller or completely absent. Some silverfish are found in the same habitats as bristletails, a few live in ant colonies, and many are familiar household insects of cracks, crevices, and dark corners (fig. 82).

Like many other soil arthropods, jumping bristletails and silverfish are very sensitive to the moisture in their microenvironments and do not survive long if the humidity drops too low. Bristletails, springtails, and other arthropods such as myriapods, diplurans, and proturans have developed special water-absorbing vesicles along with special moisture-sensitive receptors that trigger the activity of these vesicles.

The vesicles are located on the underside of the animal and are in-

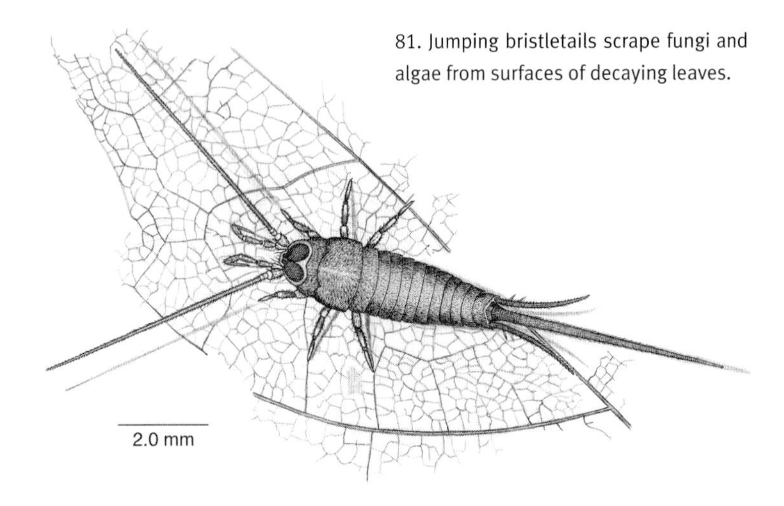

81. Jumping bristletails scrape fungi and algae from surfaces of decaying leaves.

82. Silverfish move swiftly across the ground, exploring their environment with long antennae and tail filaments. (Michael Jeffords)

flated by an increase in blood pressure whenever the humidity drops too low; they are retracted by muscles once the animal's delicate water balance is restored. Each vesicle is lined by cells that secrete a thin cuticle and are specialized for rapid uptake of water from the environment. The vesicles pop in and out of these moisture-sensitive arthropods as the humidity rises and falls, guaranteeing that the bristletail neither shrivels from lack of water nor swells from imbibing too much water.

2. Earwigs

Phylum Arthropoda
Class Insecta
Order Dermaptera
Place in food web: decomposers, detritivores, scavengers, herbivores, fungivores, algal eaters, a few predators

Impact on gardens: adversaries/allies
Size: 4–20 mm in length
Estimated number of species: 1,500

Earwigs are creatures of the night that spend their evenings abroad and their days sheltered in the soil or leaf litter. Most earwigs feed on plant debris, algae, and fungi, but a few hunt small insects and spiders. During the day they feel most at home in tight crevices and close quarters where as much of their body surfaces as possible can con-

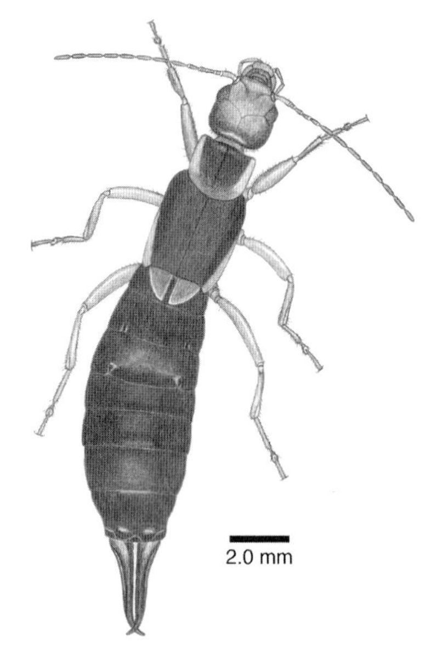

83. This female earwig will tend her clutch of eggs in an underground chamber, continually licking and rolling the eggs to remove any fungal spores that might settle on them.

tact their surroundings. This typical earwig behavior assures that they are often inaccessible to the beetles, spiders, centipedes, and birds that might feed on earwig meat. Their strong pincers or cerci offer extra protection by delivering a good pinch to any predator that ventures too close. And if a pinch is not sufficient to deter a predator, some species of earwigs such as the one illustrated here emit a foul-looking, foul-smelling brown fluid. On the dorsal surface of the third and fourth segments of their abdomens they have pairs of glands from which they can squirt this repellent as far as ten centimeters.

In late winter or early spring, an earwig mother retires to a snug chamber in the soil to lay about 50 eggs. There she stays and guards the eggs until they hatch several days later (fig. 83). Down in the soil, the young earwigs stay close together through their first four to six molts before appearing aboveground for the summer.

3. Cockroaches

People rarely appreciate how beautifully adapted cockroaches are to life in leaf litter and rotten logs, probably because people have never considered appreciating cockroaches even when they are found far from the kitchen or bathroom. The country relatives of our much maligned household cockroaches lead respectable lives as recyclers on

Phylum Arthropoda
Class Insecta
Order Blattodea
Place in food web: decomposers, detritivores, scavengers
Impact on gardens: allies
Size: 3–50 mm in length
Estimated number of species: 3,680

the forest floor. By having long legs designed to fold closely against their bodies, cockroaches can readily slip their oval, flattened bodies under logs and between layers of leaves on the forest floors. The long antennae of cockroaches swing freely in broad arcs and explore a wide area around each insect, while bristles on cerci at the tip of its abdomen detect the slightest movements of air that might portend the approach of danger.

Not all cockroaches live in fear of predators such as ground beetles, centipedes, and shrews; one of the smallest cockroaches, *Attaphila* (*Atta* = generic name for leaf-cutting ants; *phila* = love), lives in the relative security of nests built by leaf-cutting ants and survives on leftovers and handouts from the worker ants. But for many cockroaches, feeding on decaying vegetation of the forest floor can often be as dangerous as feeding on leftovers in the kitchen (fig. 84).

Some cockroaches not only live with social insects such as ants, but they also have social lives of their own. These are a special family of wood-feeding cockroaches called cryptocercids that are found only in three places on earth: the Pacific Coastal foothills of the western United States, the Appalachian Mountains of the eastern United States, and China. One big difference between cryptocercids and other cockroaches is conveyed by the family name, which translates

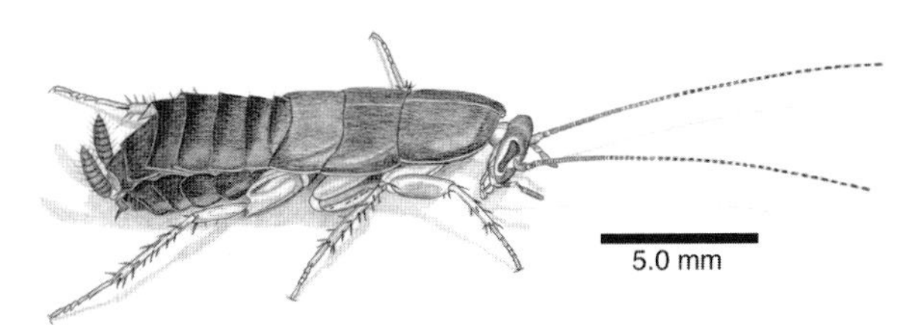

84. Cockroaches have a pair of long, slender antennae at their head end and another pair of short, stout cerci at their tail end that can detect the slightest disturbances in the leaf litter.

to "hidden cerci." The cerci of cryptocercid cockroaches are hidden under the abdomen rather than extending from the tip of the abdomen. The animals live in small colonies made up of two parents that often live for several years along with their 15 to 20 immature offspring called nymphs. Like termites, these cockroaches can survive on rotten wood that is very nutrient-poor because in their hindguts they harbor protozoa and bacteria that can enhance their digestion and nutrition. Social life in the log involves nymphs feeding on the cast-off skins of their siblings as well as on fluids from their parents' hindguts, both of which contain the indispensable microbes that are unfailingly passed from one generation of cockroaches to the next.

4. Camel Crickets and Mole Crickets

Phylum Arthropoda
Class Insecta
Order Orthoptera

Family Rhaphidophoridae (camel crickets)
Place in food web: decomposers, detritivores, scavengers, predators
Impact on gardens: allies
Size: 5–40 mm in length
Estimated number of species: 1,000

Family Gryllotalpidae (mole crickets)
Place in food web: diggers, predators, herbivores
Impact on gardens: adversaries/allies
Size: 25–50 mm in length
Estimated number of species: 50

The crickets that most of us are familiar with chirp away their evenings in treetops or among grasses. However, camel and cave crickets, the wingless members of cricket families, lead chirpless lives under stones, logs, piles of leaves, and even in the burrows of larger animals like gophers and kangaroo rats. Since they like dark, moist environments, where long, sensitive antennae are more useful than eyes, some of these crickets have lost their eyes all together. Down in the dark passageways of their homes, they may feed on plant debris, but they probably also eat smaller arthropods that come their way (fig. 85).

Mole crickets have not given up chirping and have even gone to great lengths to dig and mold underground tunnels of exceptional acoustical design. Mole crickets are clearly designed for digging. Their front legs are broad, flat, and very muscular and look remarkably like the front legs of that master burrower, the mole. As mole crickets look for earthworms, small insects, or plant roots, they bur-

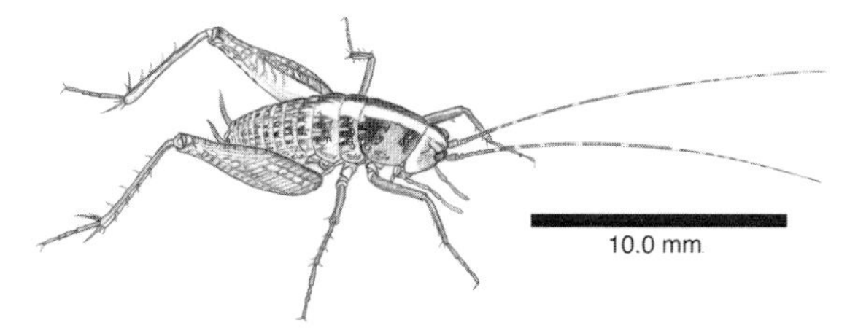

85. Most camel crickets belong to the genus *Ceuthophilus* (*ceutho* = concealment; *philus* = lover of), a name that conveys their love of dark hiding places.

row from 15 to 20 centimeters belowground. But during courtship, the male uses much of his precious energy digging a shallow burrow designed to enhance the sound of his courting chirps.

The burrow looks and sounds like a subterranean megaphone. At the bottom, the male chirps away with his head facing the narrow, dead end of the burrow and his tail facing the wider entrance to the burrow. The chirps are generated as the mole cricket raises its wings and rubs them together. His chirps are amplified and carried afield

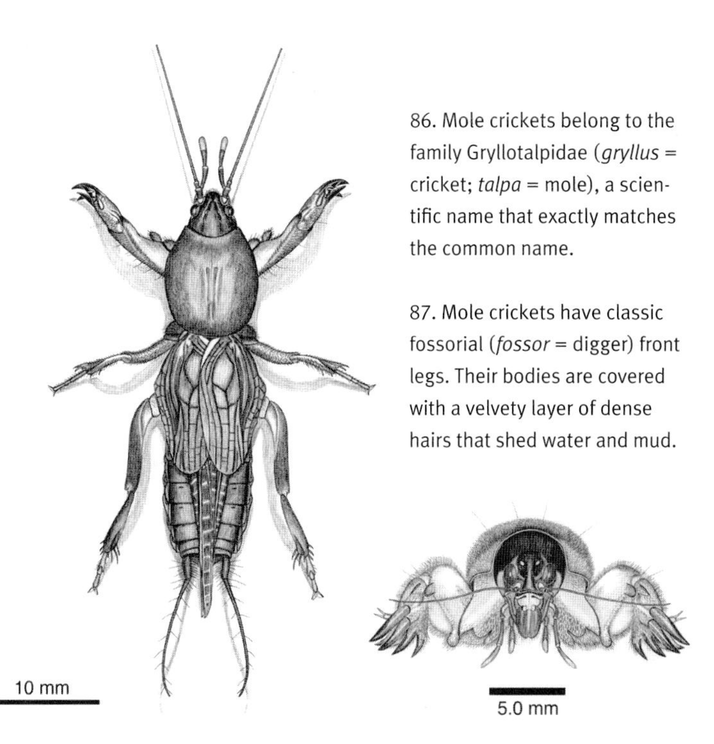

86. Mole crickets belong to the family Gryllotalpidae (*gryllus* = cricket; *talpa* = mole), a scientific name that exactly matches the common name.

87. Mole crickets have classic fossorial (*fossor* = digger) front legs. Their bodies are covered with a velvety layer of dense hairs that shed water and mud.

by his carefully constructed megaphone, advertising his whereabouts and romantic intentions. He vigorously fiddles away with one wing scraping against the other, trusting that some female will eventually hear his earnest calls (figs. 86–87).

5. Short-horned Grasshoppers

Phylum Arthropoda	**Place in food web:** herbivores
Class Insecta	**Impact on gardens:** adversaries
Order Orthoptera	**Size:** 5–115 mm in length
Family Acrididae	**Estimated number of species:** 10,000

Some insects begin life in underground chambers even though as they mature they leave the soil behind. Insects such as grasshoppers and field crickets deposit their eggs in the earth. A female grasshopper drills an inch or more into the earth with four hard prongs at the tip of her abdomen. As eggs pass from her outstretched abdomen, the female secretes a frothy material that hardens and forms a leathery case around them. A female can deposit several of these clusters of eggs, called pods, each in its own underground chamber (fig. 88).

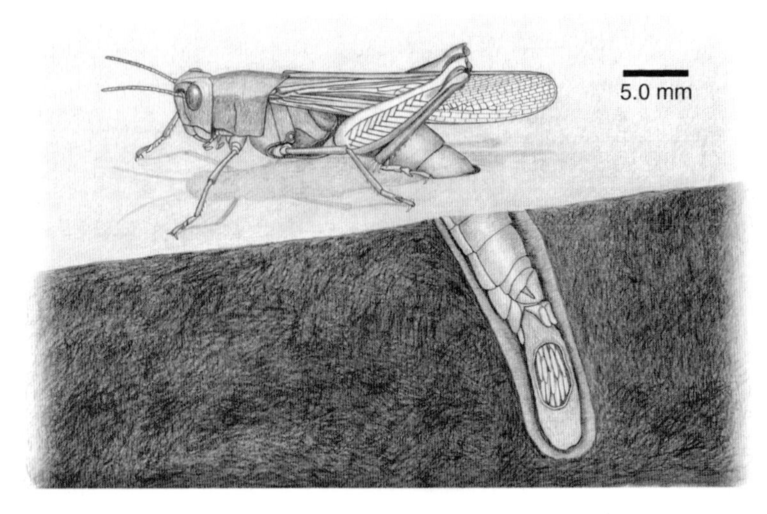

88. A mother grasshopper lays a pod of eggs beneath the soil's surface.

Anywhere from half a dozen to 150 eggs make up a grasshopper egg pod. Other insects, such as blister beetles and bee flies (see figs. 157–158), come along and lay their eggs where their own larvae can locate these caches of grasshopper eggs. These beetle and fly larvae feed on eggs throughout their larval days until metamorphosis sends them aboveground to feed on flowers and leaves (plate 24).

Soil provides a nursery for many other creatures—small and large. Turtles, snakes, lizards, and salamanders often shelter their eggs in underground chambers. Some birds such as puffins, swallows, motmots, and kingfishers lay their eggs in burrows. Burrowing mammals give birth in the quiet and safety of their underground dens. The earth shelters these eggs and young animals from the elements of weather and ravaging of predators.

6. Termites

Phylum Arthropoda
Class Insecta
Order Isoptera
Place in food web: diggers, detritivores, decomposers, some fungivores

Impact on gardens: absent
Size: 6–7mm in length
Estimated number of species: 2,300

While the earthworm is the chief tiller of soils at temperate latitudes, in warmer and more tropical climes the termite is undoubtedly the principal purveyor of leaves, twigs, and wood to the underworld. So efficient are these tropical insects at clearing plant litter from the soil surface and so efficient are the soil bacteria and fungi of the tropics at further degrading the plant matter that humus never really has a chance to accumulate, much to the detriment of tropical agriculture (fig. 89).

Tropical soils are notoriously poor for agriculture even though they can nurture lush vegetation and magnificent trees. Termites, fungi, and bacteria quickly recycle nutrients of the forest—from plants back to soil. Practically all the organic matter of a tropical forest is tied up in the living plants of the forest; organic matter quickly disappears from the soil once the forest is gone. In a patch of forest where all the trees have been cleared for crops, rains quickly leach the few remaining nutrients of the soil beyond the reach of the crops' roots. After

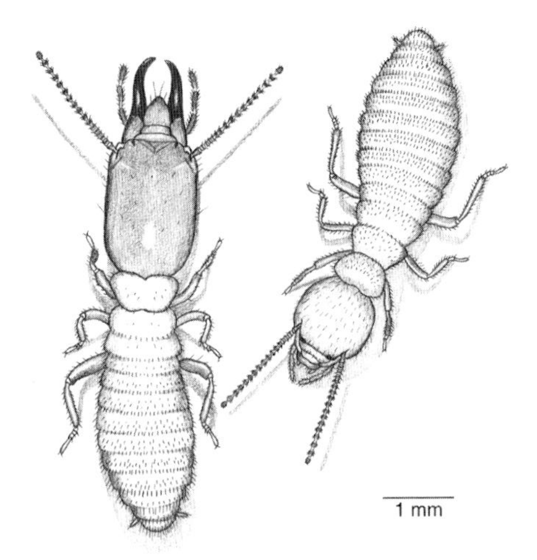

89. Workers (right) and soldiers share the work in a termite colony. Soldiers protect the colony from intruders, and workers see to the foraging and housekeeping.

only three or four years the nutrients of the tropical soil have been exhausted, leaving the land practically useless for agriculture.

Termites that feed on dead wood and grass, some of the most indigestible and nonnutritious of foods, must consume vast quantities of this carbon-rich, nitrogen-poor food in order to survive and grow. In fact, termites of the African savannahs eat more grass than all the grazing mammals, such as wildebeests, zebras, and gazelles, put together.

However, tucked away at the rear end of every termite's gut are colonies of microbes that share meals with their host. So many bacteria and protozoa may dwell in the gut that they often make up one third of the insect's weight. The microbes provide enzymes that supplement those produced by cells of the gut. They also use up some of the excess carbon and help concentrate the nitrogen that the termites need to grow. You might ask where the bacteria and protozoa get the supply of nitrogen that they need for their own growth among all this nitrogen-poor cellulose. In return for their contributions to the good digestion of the termite, their host provides them with well-chewed wood, a home, a share of the digested wood, and a rich source of nitrogen in the form of its recycled urine.

An insect's urine-excreting equivalents of kidneys are called Malpighian tubules. In termites these tubules lie against the outer surface of the gut; rather than excreting nitrogen with the urine, these tubules recycle nitrogen back into the gut. There, certain bacteria convert nitrogen in the form of uric acid into forms that are nutritious to both microbes and termites.

Termites of a colony constantly groom one another and share these microbes. Any termite that fails to get its share of these creatures or in some way loses them from its gut quickly starves to death, even when surrounded by vast supplies of plant materials. Every year, termites in the dry grasslands of Arizona clear up to 92 percent of the approximately 200 kilograms of plant litter found on an acre of ground; the gut microbes can take much of the credit for this feat.

In the tropics, termites have gone a step further in recruiting help to improve the nutrient content of their diet. They have actually become farmers of fungi. In large chambers of their colonies, termites cultivate fungi on the droppings that accumulate after pieces of wood and grass have passed through their guts the first time around. The wood and grass that pass through the second time around have been predigested not only by microbes but also by the fungi in the termites' garden. Most of the carbon from the wood and grass has been reduced to methane gas, and nitrogen from the original plant litter is now highly concentrated in the nutritious fungi. The termites feast from these gardens, extracting a phenomenal amount of nutrition from what starts out as a poor diet for any creature.

Whenever and wherever termites colonize aboveground, they carry soil with them. Whether to establish an outpost in a nearby tree or to invade the upper floors of a wooden house, termites are able to advance into the alien environment aboveground by building their tunnels of soil wherever they go. These foraging tunnels are constantly being remodeled, abandoned, and extended as new plant litter is discovered on the surface of the ground. In these tunnels they bring with them the familiar darkness and dampness of the underground.

The familiar termites of temperate latitudes lead retiring lives underground. Only when they construct earthen tunnels and venture aboveground into our homes and buildings do we take notice of them. Their relatives in the tropics and deserts, however, build imposing edifices of soil that dot the landscape like gnarled and ghostly figures. To raise these mounds of soil aboveground, and to construct their labyrinthine galleries underground, the termites are constantly moving soil about.

Countless tons of soil are mixed by termites, especially the large

90. Termite mounds dominate many savannahs of Africa, Australia, and South America.

colonies of mound-building termites that are found in Africa, Australia, and South America. Their mounds can take on familiar or outlandish forms (fig. 90). Turrets, spires, obelisks, pyramids, or mesas may rise 3 meters above the ground and extend 15 meters in diameter. A network of tunnels radiates from each mound for about 50 meters in every direction. Frequently as many as 30 such mounds may occupy an acre, and more than 5,000 tons may be moved aboveground during the construction of these mounds. These monumental mounds are a testament to the prowess of these termites as earth movers and architects.

7. Thrips

Phylum Arthropoda
Class Insecta
Order Thysanoptera
Place in food web: fungivores, algal eaters, predators, herbivores

Impact on gardens: adversaries/allies
Size: 0.5–2 mm in length
Estimated number of species: 4,700

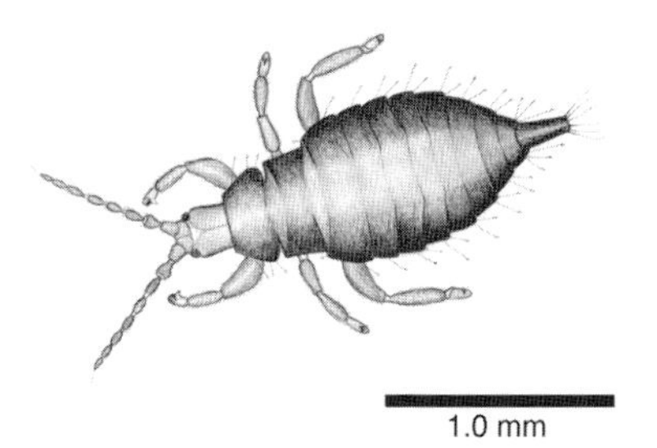

91. The group to which thrips belong is the order Thysanoptera (*thysano* = fringe; *ptera* = wings). Most thrips have fringed wings and live aboveground on plants; but all immature thrips and some of the soil thrips, such as this one, are completely wingless.

Thrips do not seem to do things the way other insects do. A thrips' mouthparts look like a cone-shaped beak, and they are positioned at the base of the head rather than at the front. While other insects have two symmetric jaws or mandibles, thrips have only one mandible, and for some unexplained reason this mandible is always a left one. The mandible works along with a second, symmetrical pair of mouthparts, or maxillae, to rasp and punch food while the cone of tissue that surrounds the mandible and two maxillae is used to suck the fluids from the food (fig. 91).

The legs of thrips can have one or two claws; but at the very tips of the legs are rounded bladders on which thrips actually walk. These bladders do not look like they provide a very substantial grip on surfaces over which the thrips tread, but they have apparently not fared any the worse for having such strange feet. As they walk across a fallen leaf, they give the impression of ballerinas gliding past on tiptoe.

Thrips of different sizes, shapes, and families live in the leaf litter and soil. Many are spore feeders or fungus feeders; others survive as predators of eggs, mites, springtails, and smaller insects. Before young larval thrips grow their wings, they settle down, stop eating, and stop walking. So many tissues are undergoing remodeling at this time that the young thrips remain immobile as old tissues break down and new tissues take their places. This stage in the life of a thrips is often called the "pupa," but it is not like the pupal stage of other insects. In young thrips, wing buds form on the outside at the very beginning of development; in insects with true pupae, wing buds first form inside the bodies of larvae, and only later do the future wings move from inside to out as the insect transforms to a pupa.

No other insects feed quite like thrips, walk like thrips, or develop like thrips. Thrips clearly belong to a unique order of insects.

8. Big-eyed Bugs and Burrower Bugs

Phylum Arthropoda
Class Insecta
Order Hemiptera

Family Lygaeidae
Subfamily Geocorinae (big-eyed bugs)
Place in food web: predators
Impact on gardens: allies
Size: 3–5 mm in length
Estimated number of species: 200

Family Cydnidae (burrower bugs)
Place in food web: herbivores, diggers
Impact on gardens: absent
Size: 5–7 mm in length
Estimated number of species: 300

As ground-dwelling predators of a mainly seed-eating, plant-dwelling family of bugs, these geocorine (*geo* = earth; *coris* = bug) bugs have accepted a lifestyle very different from that of most other family members (fig. 92).

With broad heads and large eyes, these bugs look well-suited for surveying their hunting territories among the grassroots and grass litter of meadows. Big-eyed bugs typically scurry over the ground beneath tall stalks of grasses until they spot prey that they can easily outrun. Their prey often happen to be lethargic thrips or juicy springtails. With mouthparts molded into a long, pointed beak that is used for sucking rather than for chewing, a bug's feeding is a very different enterprise than it is for us creatures with jaws.

There are two channels in the long beak of the big-eyed bug just as there are in the beaks of all true bugs; one channel carries saliva and enzymes into the body of its prey and the other carries partially digested body fluids of the prey back to the bug's gut. Massive muscles that fill the bug's broad head pump the fluids into the esophagus and gut for final digestion of the prey.

As it roams over the top of the soil, the big-eyed bug crosses paths with decomposers, diggers, and a wide variety of prey; but even this swift and sharp-eyed bug has to watch out for predators like ground beetles and rove beetles that can outweigh, outrun, and outmaneuver it.

Burrower bugs are clearly designed for digging among plant roots. The segments of its front legs are broad, and fringed with strong spines for scooping. Some bugs dig as deep as a foot and a half (45 cen-

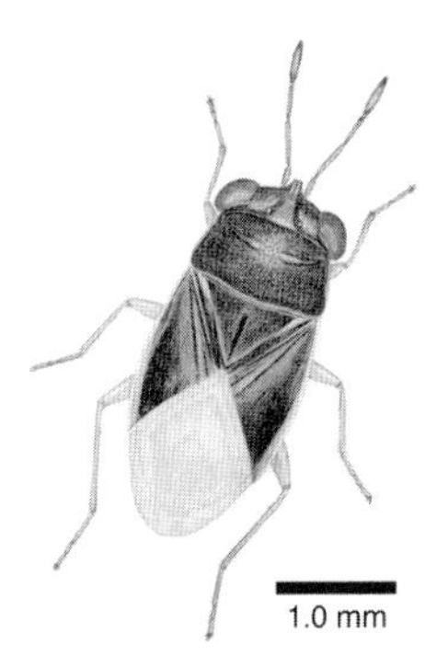

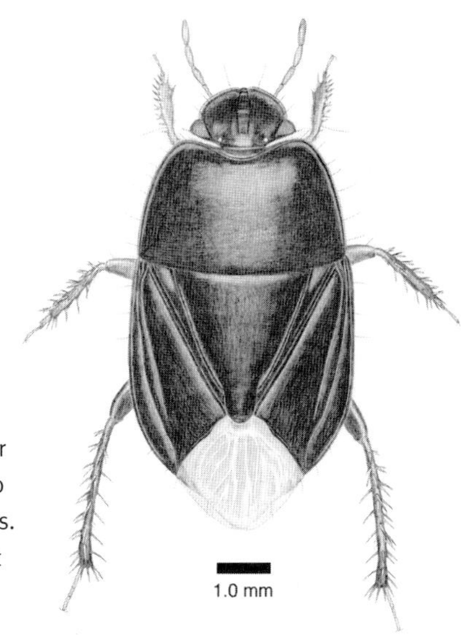

92–93. Big-eyed bugs and burrower bugs share the same habitat but do not compete for the same resources. Big-eyed bugs (92, above) are swift hunters, and burrower bugs (93, right) dig for roots.

timeters) into the ground looking for roots. Here in a chamber among the roots, a female burrower bug lays 30 to 150 eggs and steadfastly stays with them until they hatch. Like the spiderlings of a mother wolf spider, the young burrower bugs crawl onto their mother as soon as they hatch. They stay with her for several days as she nurses them with drops of fluid from her anus that they avidly gobble up. For the first few days of their lives, their mother's fluid is their only nourishment. Only young bugs nursed by their mothers survive to feed on plant roots. What the mother bug passes on to her children are bacteria that live in the gut of every healthy burrower bug. These bacteria produce vitamins and other nutrients that supplement the bugs' diet of root juice (fig. 93).

9. Aphids, Phylloxerans, and Coccoids

Scale insects (coccoids), phylloxerans, and aphids are closely related insects. They all have beaks for sucking plant juices, some have glands on the surfaces of their bodies that secrete waxy fibers or powder, many females give birth to living young, and some females give birth without even mating.

Most aphids and phylloxerans live aboveground on leaves and stems, but there are a few that pass their days underground. In fact,

Phylum Arthropoda
Class Insecta
Order Hemiptera

Aphids and Phylloxerans
Families Aphididae (aphids) and **Phylloxeridae** (phylloxerans)
Place in food web: herbivores, fungivores
Impact on gardens: adversaries
Size: 1–8 mm in length
Estimated number of species: 3,500 (aphids); 70 (phylloxerans)

Coccoids
Superfamily Coccoidea
Place in food web: herbivores, fungivores
Impact on gardens: adversaries
Size: 1–4 mm in length
Estimated number of species: 6,000

the name phylloxera (*phyllo* = leaf; *xero* = dry) refers to the effect that these insects have on plants aboveground. Those that reside in leaf litter sometimes poke their beaks into the abundant filaments of fungi found on decaying leaves and feed on fungal juices. Aphids and phylloxerans are not equipped for digging, but still they can be found deep below the layer of fallen leaves, where they feed on root sap (fig. 94).

Ants and aphids have established mutually beneficial partnerships in which ants take responsibility for the welfare of the aphids, and the aphids reward their masters with their sugary droppings, commonly known as honeydew. There are ants that dutifully carry eggs of root aphids into their nests every autumn and in the spring they carry the newborn aphids to newly sprouted roots. Throughout the growing season the ants carry the aphids from root to root to assure a steady supply of sweet honeydew.

Scale insects, also known as coccoids, are a very diverse group of plant feeders, so diverse in their forms that they represent a superfamily, Coccoidea, that has been divided into 16 families. Some scale insects have only females; in those species the mother gives birth with-

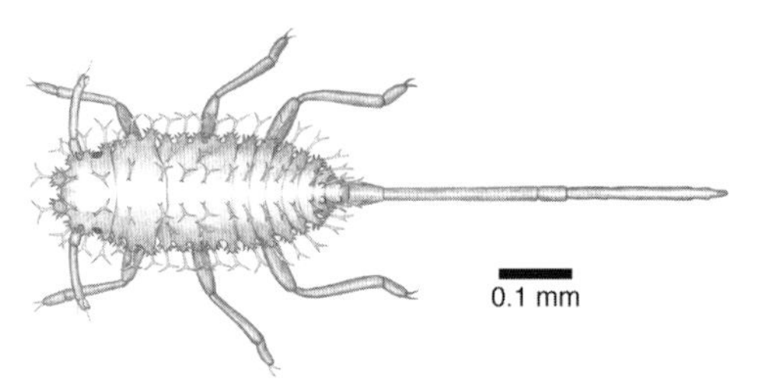

94. This phylloxera from the soil of a maple grove has a beak that extends under its body and well beyond the tip of its abdomen.

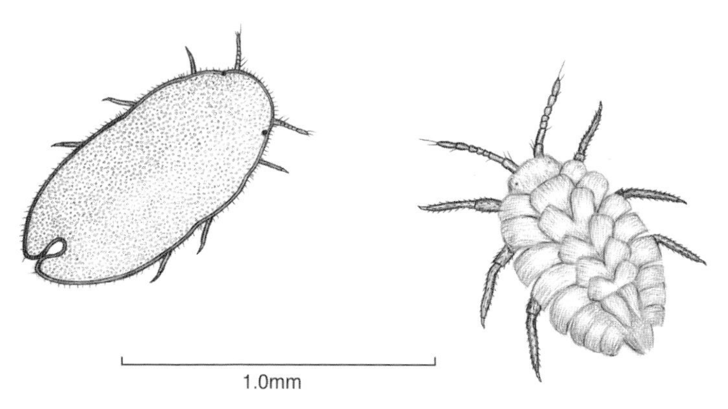

95. These two coccoids from the soil of an American forest are representative of the great diversity of forms among the coccoids. An ensign scale insect (right) that secretes waxy plates from its skin and a very flat scale insect (left) can each have 1,000 to 2,000 siblings.

out ever mating. In those species that do have males, the males have wings and legs but no mouthparts with which to feed; their lives as adults are very ephemeral. They quickly mate and live only one or two days. The females are the ones with mouthparts, even though they are wingless and usually legless as well.

As newly hatched nymphs, scale insects have legs and are nomadic; but at their next molt these legs are lost and the nymph settles down to feed in one spot. If the nymph is a female, it usually stays there for the rest of its life. A male nymph will usually fly off once it becomes an adult. Many members of the coccoid families feed on parts of plants found aboveground. On twigs and leaves, armored scales, wax scales, tortoise scales, pit scales, and soft scales are totally sedentary as females and reside under a scalelike covering secreted from glands on the body surface. A good number of coccoids, however, have settled down to life in the underground, feeding on the sap of roots and fungi (fig. 95).

Female members of one coccoid family have smooth metallic gold or pearly shells, feed on rootlets, and are so round that they are referred to as ground pearls. In parts of Africa and on St. Vincent in the Lesser Antilles, these hard, rotund insects are strung as beads on necklaces. Members of another family, the ensign scales, are covered with white, waxy plates and lumber around rootlets and filaments of fungi. Some subterranean coccoids can be so flat that it is a wonder there is space enough for heart, nerve cord, and other essential organs between their backs and their bellies. A variety of shapes and forms serve quite well for coccoids that live in the soil.

10. Cicadas and Rhipicerid Beetles

Phylum Arthropoda
Class Insecta

Cicadas
Order Hemiptera
Family Cicadidae
Place in food web: herbivores, diggers
Impact on gardens: absent
Size: 25–50 mm in length
Estimated number of species: 1,500

Rhipicerid Beetles
Order Coleoptera
Family Rhipiceridae
Place in food web: parasites
Impact on gardens: absent
Size: 21–24 mm in length
Estimated number of species: 52

Witnessing an emergence of red-eyed cicadas in late May and early June is an experience never to be forgotten, and one that occurs only once every 13 or 17 years in any one location. As the nymphs of these periodical cicadas emerge en masse after spending over a decade belowground, they leave the soil surface perforated with their emergence holes (fig. 96). The nymphs can be very abundant in the soil; twenty thousand of these quarter-inch-wide holes were counted under a single apple tree in Indiana. The millions of male cicadas that soon fill the trees raise a deafening racket as they vie for the at-

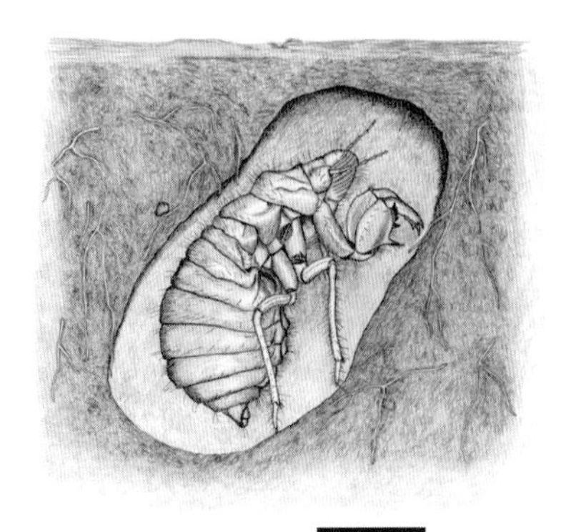

96. After spending most of its life in the soil, a cicada nymph sheds the skin of its youth, finally unveiling glistening wings and lovely adult colors.

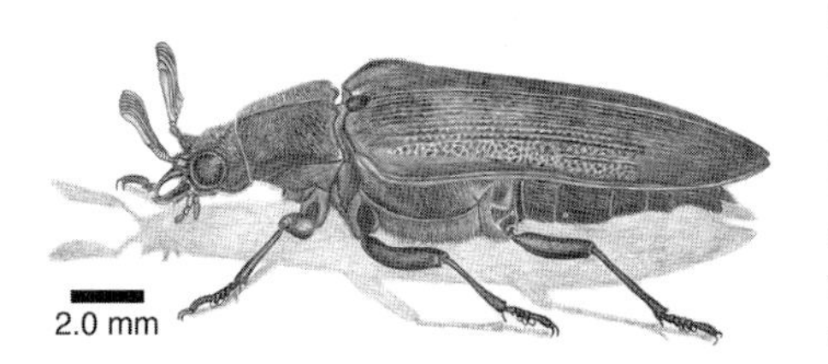

97. Parasitic larvae of rhipicerid beetles use nymphs of cicadas as their hosts. These beetles do not compete with many predators and parasites for this well-hidden resource.

tentions of the noiseless females. After mating in the treetops, the females insert their eggs in twigs by slicing through the bark with their sharp ovipositors (*ovum* = egg; *positum* = deposited) (plate 25).

Within a few days, tiny cicada nymphs emerge from the eggs, falling many feet to the ground and beginning their long journeys among the roots. The nymphs immediately tap the roots with their sharp beaks and feed on sugary sap until one day, many years later, some unknown cues tell the nymphs that the time has come to leave the soil behind.

The larger cicadas that emerge every year during the dog days of July and August have brown eyes and shorter life cycles than the periodical cicadas. They are known to have life cycles that last as long as seven years, and the shortest time that they spend underground is still four years.

The larvae of one family of beetles, the Rhipiceridae, specialize as parasites of cicada nymphs. The name for the family (*rhip* = fan; *ceri* = horn) comes from the beautiful fan-shaped antennae of the adult beetles. They use their antennae to find trees where cicadas have recently laid their eggs; they then lay large numbers of their own eggs on the trunks of these trees. The tiny, but very active, beetle larvae that hatch from the eggs are able to wend their way down through the pores and cracks of the soil until they find cicada nymphs. Even deep beneath the soil surface cicada nymphs must contend with parasitic insects. Once the fast and agile rhipicerid larva finds a nymph, it becomes sedentary and obese, leading the decadent life of a typical parasite (fig. 97).

11. Rove Beetles and Ground Beetles

In North America, rove beetles can claim to be the largest family of beetles. Most match the soil in color, but some found on mushrooms can be attractively iridescent.

Phylum Arthropoda
Class Insecta
Order Coleoptera

Rove Beetles
Family Staphylinidae
Place in food web: predators, diggers
Impact on gardens: allies
Size: 0.7–25 mm in length
Estimated number of species: 50,000

Ground Beetles
Family Carabidae
Place in food web: predators, diggers
Impact on gardens: allies
Size: 1.5–35 mm in length
Estimated number of species: 30,000

Rove beetles can be intimidating, especially the larger ones, which are about an inch long. One particularly fierce-looking British rove beetle goes by the name of Devil's coach horse. The short wing covers, or elytra, of a rove beetle give its long abdomen a great deal of freedom to maneuver in the tight spaces of the soil. Rove beetles are swift and often dash about with their abdomens threateningly raised and poised like a scorpion's tail. The raised tail is an empty threat, but the long, sickle-shaped jaws at the other end can deliver a good bite (figs. 98–100).

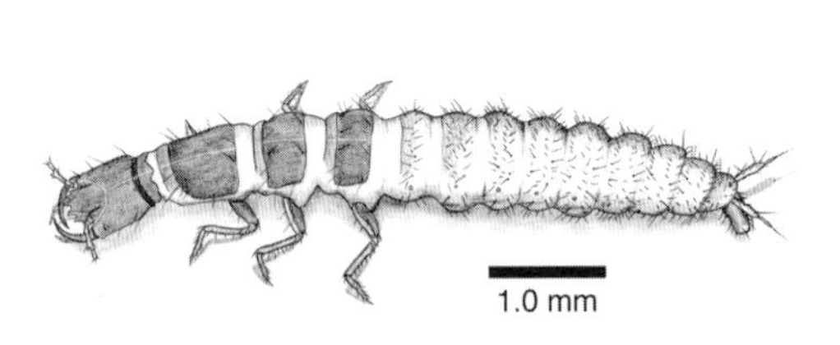

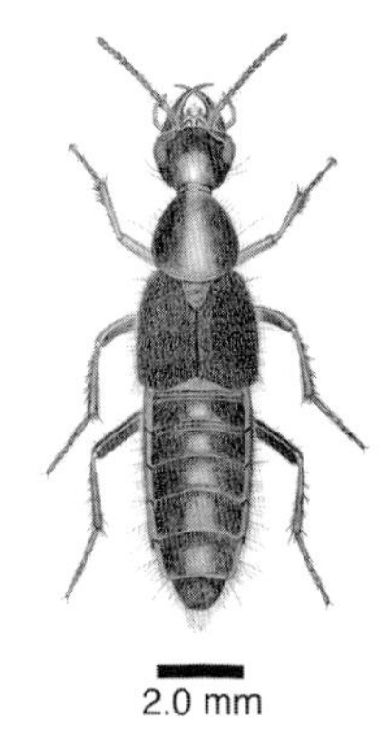

98–99. Even after metamorphosis, the fierce demeanor of larval rove beetles (98, above) is maintained by the adult beetles (99, right).

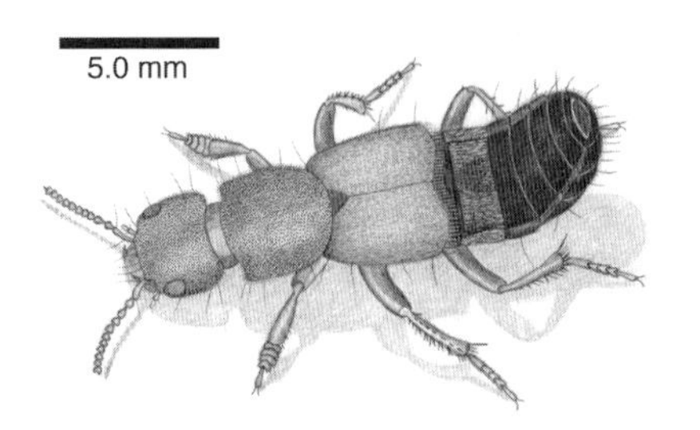

100. A rove beetle indicates its irritation by curling its abdomen above its back in a threatening pose.

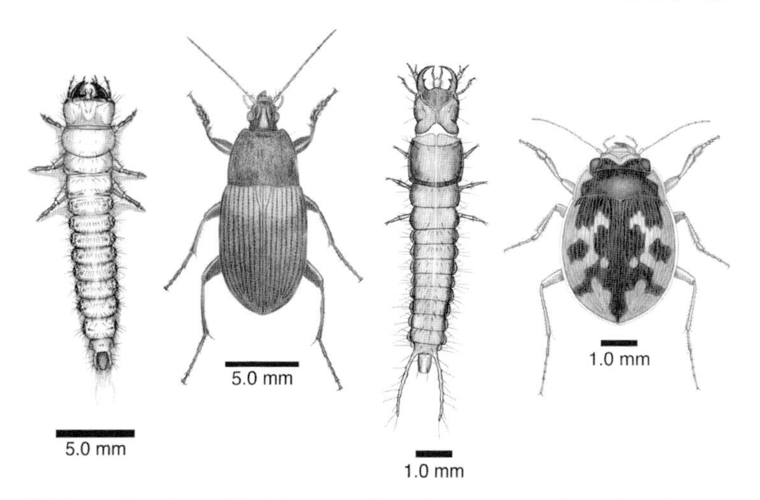

101. Ground beetles and their larvae are fierce, fast-moving, and usually quite common. The larvae blend with the soil, but the adults can be colorful, iridescent, and flashy.

Rove beetle larvae, like their parents, have powerful jaws, speed, and agility. They also live in the same habitats and prey on some of the same animals as their parents. They are often confused with the larvae of ground beetles. However, while each leg of a rove beetle larva ends in a single claw, ground beetle larvae have a pair of claws on each of their six legs and a pair of what look like tails at the tip of their abdomens. These structures are known as urogomphi (*uro* = tail; *gomphi* = tooth) and can either be jointed where they meet the abdomen, as in rove beetle larvae,or nonjointed, as in ground beetle larvae.

Each ground beetle usually consumes well over its own weight in food in a day—sometimes even two and a half times its own weight. Once its mandibles crush its prey, the beetle regurgitates digestive juices so that digestion of its prey is well along by the time the beetle finally swallows the now tenderized and liquefied meal.

A series of field experiments in Britain illustrates how effective ground beetles and rove beetles are as predators, as well as how pesticide application can often be counterproductive. These experiments involved application of insecticides to cabbage fields to control damage from maggots of the cabbage root fly (fig. 162). Ironically, fields treated with pesticides suffered more damage than fields that were left untreated. A closer look at the situation revealed that over 30 spe-

102. The burrowing ground beetle is a tiny predator of rove beetles.

cies of beetles were eating the eggs, larvae, and pupae of the cabbage root fly, but the pesticides were killing these beetles. If left to their own devices, the beetles were far more effective than pesticides in keeping the root-maggot flies under control (fig. 101).

While most ground beetles run down their prey and hunt on the soil's surface and in the leaf litter, some of the smaller ground beetles, known as burrowing ground beetles, actually live in burrows rather than under stones or logs. These sleek and compact little beetles have large front legs clearly designed for digging after the small rove beetles and mud-loving beetles that make up a good portion of their diet (fig. 102).

12. Tiger Beetles

Phylum Arthropoda	**Place in food web:** diggers, predators
Class Insecta	**Impact on gardens:** allies
Order Coleoptera	**Size:** 6–40 mm in length (mature larvae)
Family Carabidae	
Subfamily Cicindelinae	**Estimated number of species:** 1,300

Tiger beetle larvae start out living in burrows that are about six inches (15 centimeters) deep; but by the time they are full grown, they have excavated vertical burrows down into the subsoil as deep as a foot and a half (45 centimeters). The burrows are often close together in places where the soil is exposed and vegetation is sparse, not because the larvae are social but because their mother generally left several eggs in one spot.

1. These sandstone rocks in the hills of southern Illinois have been weathered and sculpted by thousands of years of rain, wind, and snow. (Michael Jeffords)

2. Lichens and mosses are the first vegetation to colonize barren rocks. Once these organisms have settled down on a rock, such as the one above, they can remain for hundreds, sometimes thousands, of years. (Michael Jeffords)

3. This crytobiotic soil in Arches National Park, Utah, represents an early stage in soil formation. (Michael Jeffords)

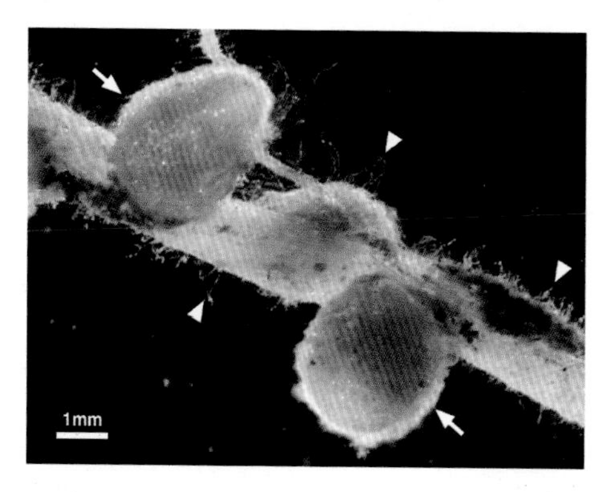

4. The nodules on the roots of legumes (arrows) form as a collaborative effort of rhizobial bacteria and cells of the plant that they enter via root hairs (arrowheads). Because the bacterial dinitrogenase enzyme that converts dinitrogen gas to ammonia works best in the absence of oxygen, the root cells have devised a strategy for keeping oxygen out of the way of the dinitrogenase. The root cells produce a red protein similar to the red blood protein hemoglobin that binds and carries oxygen from our lungs and through blood vessels to every cell of our bodies. The red protein that legumes produce is called leghemoglobin; it is what imparts the red color to these nodules of a clover root and sequesters oxygen from dinitrogenase.

5. Mushrooms are the colorful fruiting bodies of certain fungi. When weather conditions are just right, mushrooms arise from the miles of fungal filaments that live in the soil and leaf litter of the forest. (Michael Jeffords)

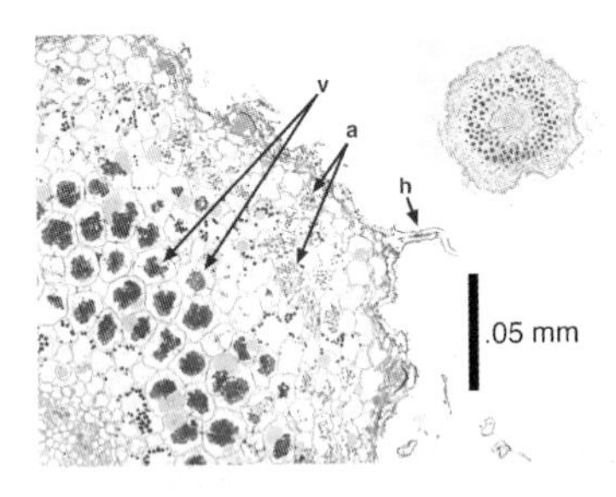

6. Endomycorrhizal (vesicular-arbuscular mycorrhizae, VAM) fungi associate with roots of wild orchids. The filaments enter a root through its root hairs (h) and then form vesicles (v) and arbuscles (a) within individual root cells.

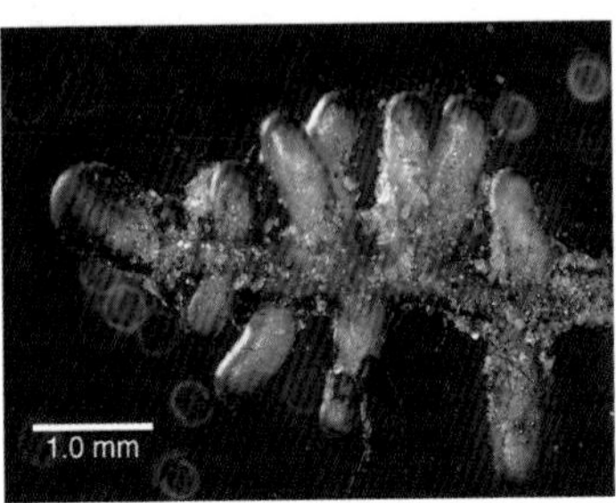

7. Ectomycorrhizal fungi associate with tree roots such as this root of a beech. The fungal filaments have been stained blue to distinguish them from cells of the beech root. Most filaments form a thick mantle around the surface of the root (arrow, right image); some filaments enter the interior of the root by passing between root cells but never entering individual root cells.

8. Plant roots contribute to soil formation by physically cracking rocks. Roots also chemically alter rocks as the carbon dioxide they produce combines with water in the soil and forms corrosive carbonic acid. (Michael Jeffords)

9. Tall saguaro cacti and short mesquite shrubs share the Arizona desert. The shallow roots of saguaro and the deep roots of mesquite do not compete for the sparse water of desert soil. (Michael Jeffords)

10. Mustard plants grow better in soil to which dead plant matter has been added (pots 5 and 6, topsoil and subsoil, respectively) than they do in soil to which no plant matter has been added (pots 2 and 4, topsoil and subsoil, respectively). However, they do not grow as large in soil where rye plants had previously grown (pots 1 and 3, topsoil and subsoil, respectively). They grow best in soil that contains both dead plant matter and a greater number of living soil creatures (pot 5, topsoil). (Michael Jeffords)

11. Death and decay, life and rebirth flow together among the decaying leaves and logs of this forest floor in the Adirondack Mountains of New York. (Michael Jeffords)

12. White rot fungi form mushrooms with sexual spores and produce an array of enzymes that digest the cellulose, hemicellulose, and lignin of wood. (Michael Jeffords)

13. Lichens are among those very few creatures that can colonize bare rocks. The color patterns of many insects, such as the grasshopper in the lower photograph, beautifully blend with the forms and colors of lichens in their habitats. (Michael Jeffords)

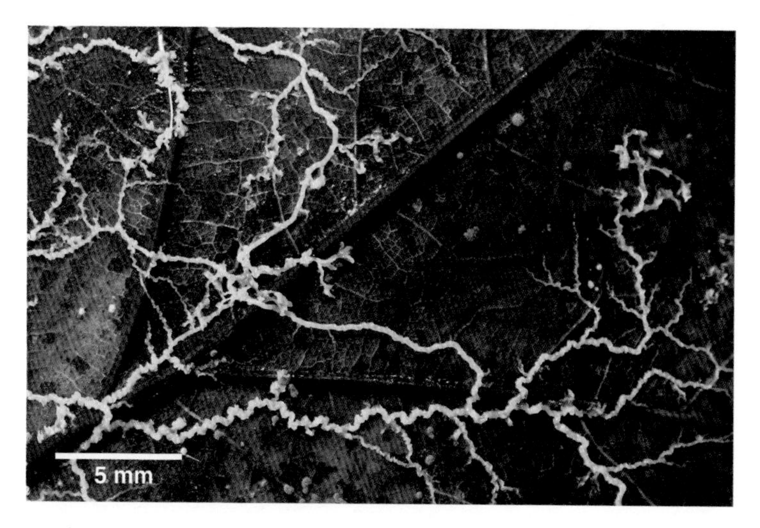

14. The bright yellow plasmodium of slime molds changes its shape within minutes as it oozes over leaves and logs. The slime mold *Physarum* engulfs most of the microbes that lie in its path. (Michael Jeffords)

15. These snails have gathered on a moist log in the southern Appalachians. (Michael Jeffords)

16. Polydesmid millipedes (order Polydesmida) are blind, often brightly colored millipedes that can secrete cyanide. (Michael Jeffords)

17. *Narceus americanus* (order Spirobolida) is a common North American millipede that secretes foul-smelling benzoquinone from glands on the sides of its body segments. (Michael Jeffords)

18, 19. Both true scorpions (18, top) and windscorpions (19, bottom) live in desert soils. Windscorpions can bite with their huge jaws, and true scorpions can sting with their long tails. (scorpion, Michael Jeffords; windscorpion, Joe McDonald/Visuals Unlimited)

20. The largest species of whipscorpion is the vinegaroon. It does not sting but sprays a vinegar-scented mist at prey and predators. (Gerold and Cynthia Merker/ Visuals Unlimited)

21. The long, whiplike first pair of legs of the tailless whipscorpion give it its name. (Ken Lucas/Visuals Unlimited)

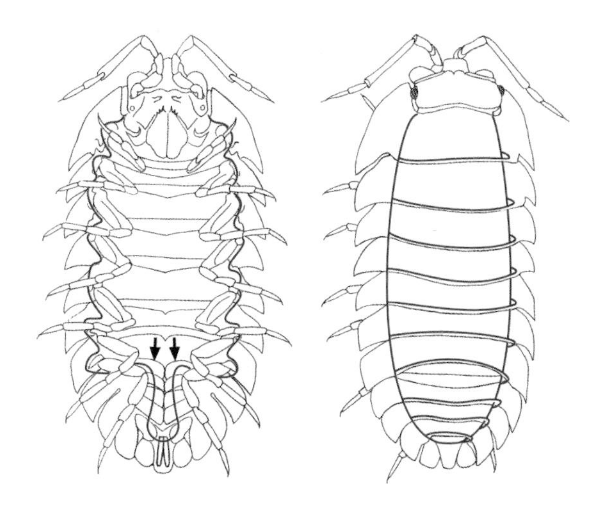

22. Around the outside edge of each segment as well as between segments (red lines), the recycled urine of a woodlouse continually flows; its four pairs of pleopods are indicated with arrows.

23. Crawfish frogs sometimes move into the spacious tunnels dug by crayfish. (Michael Jeffords)

24. Certain species of mother blister beetles (family Meloidae) have specialized in finding grasshopper eggs as food for their larvae. (Michael Jeffords)

25. Periodical cicadas spend only a few days of their 13- or 17-year lives aboveground. (Michael Jeffords)

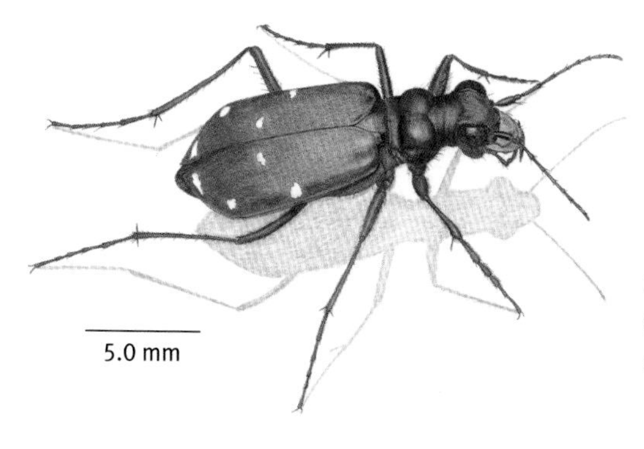

26. Tiger beetles show off their colorful iridescence on sunny paths and sand bars.

27. A male firefly glides a short distance above the ground and flashes to female fireflies down below. (Michael Jeffords)

28. Soldier beetles can be very common visitors to flowers such as goldenrod and milkweed. Some flowers can have several beetles visiting at the same time.

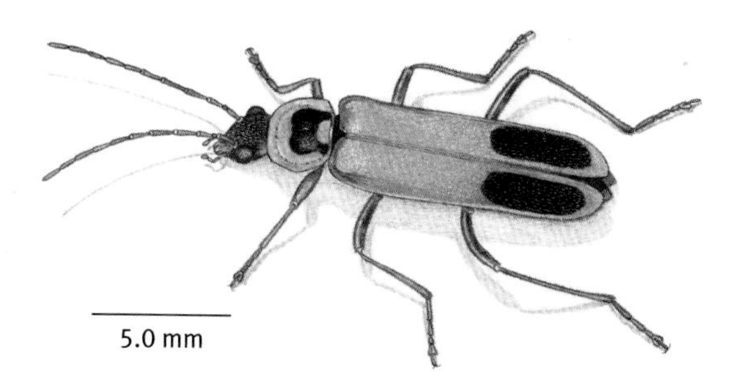

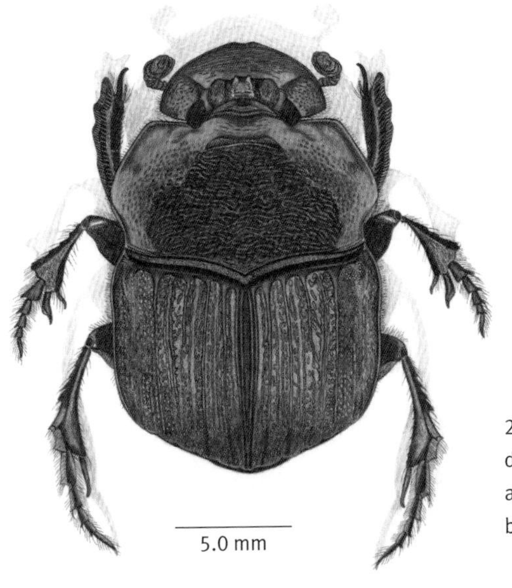

29. Many dung beetles are dark and earth-colored, but a few are known for their brilliant colors.

30. This passalid beetle inspects the surface of a rotting log as it transports a group of soil mites on the top of its head. (Michael Jeffords)

31. The imposing jaws of stag beetles can be deceiving; stag beetles with small jaws usually deliver harder pinches than those with the larger jaws. (Michael Jeffords)

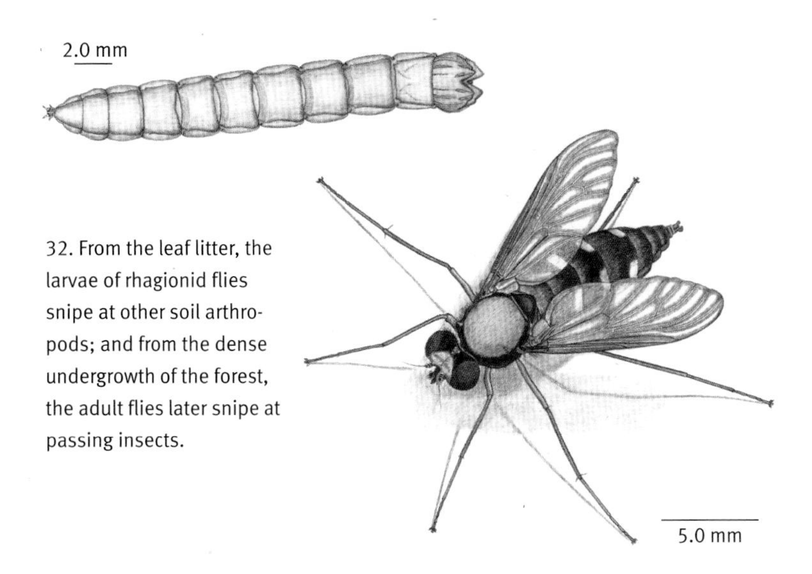

32. From the leaf litter, the larvae of rhagionid flies snipe at other soil arthropods; and from the dense undergrowth of the forest, the adult flies later snipe at passing insects.

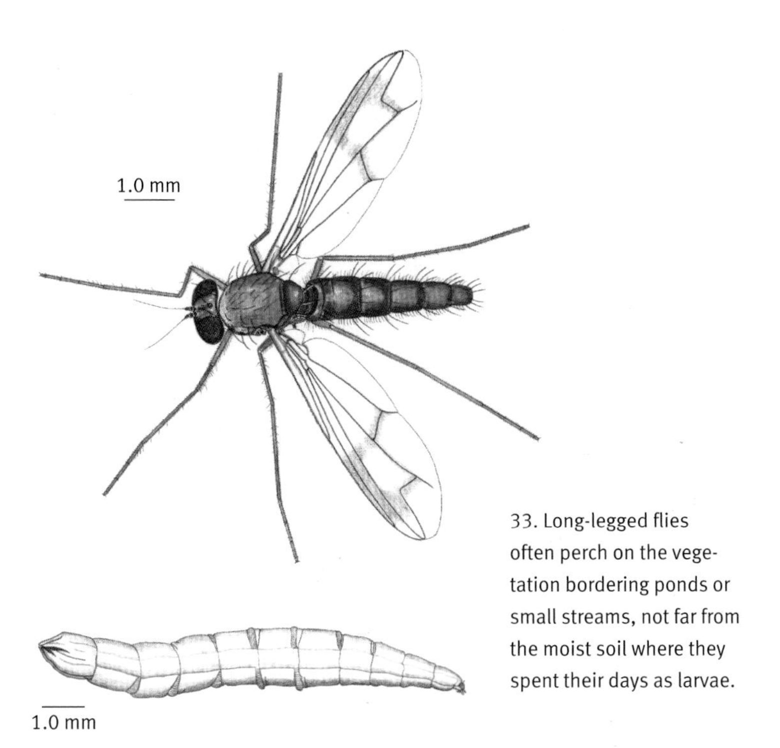

33. Long-legged flies often perch on the vegetation bordering ponds or small streams, not far from the moist soil where they spent their days as larvae.

34. Spider wasps are long-legged, usually dark-winged wasps that will tackle spiders as large as tarantulas. (Rick Poley/ Visuals Unlimited)

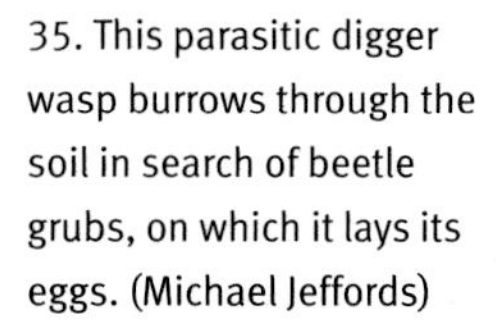
35. This parasitic digger wasp burrows through the soil in search of beetle grubs, on which it lays its eggs. (Michael Jeffords)

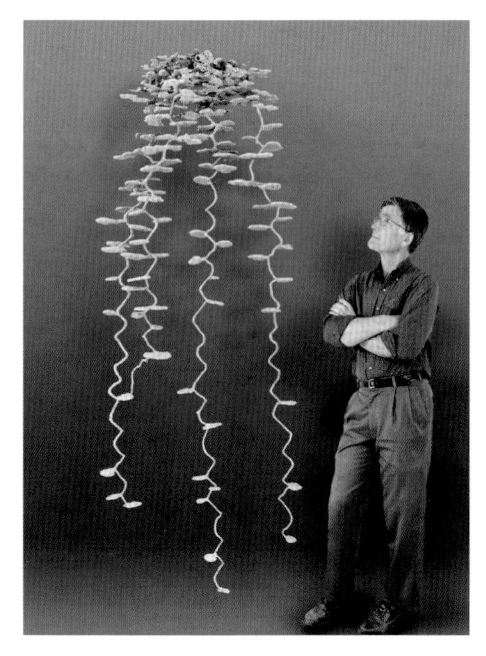

36. An authority on ants, Dr. Walter Tschinkel, poses with the cast of a harvester ant colony that he carefully excavated after pouring plaster down the colony's burrow. Five thousand ants dug these tunnels and chambers in only five days. (photo taken by Charles F. Badland)

37. The spotted salamander is one of 31 mole salamanders that live only in North America. (Michael Jeffords)

38. The colorful eft is the terrestrial phase of the aquatic newt. (Michael Jeffords)

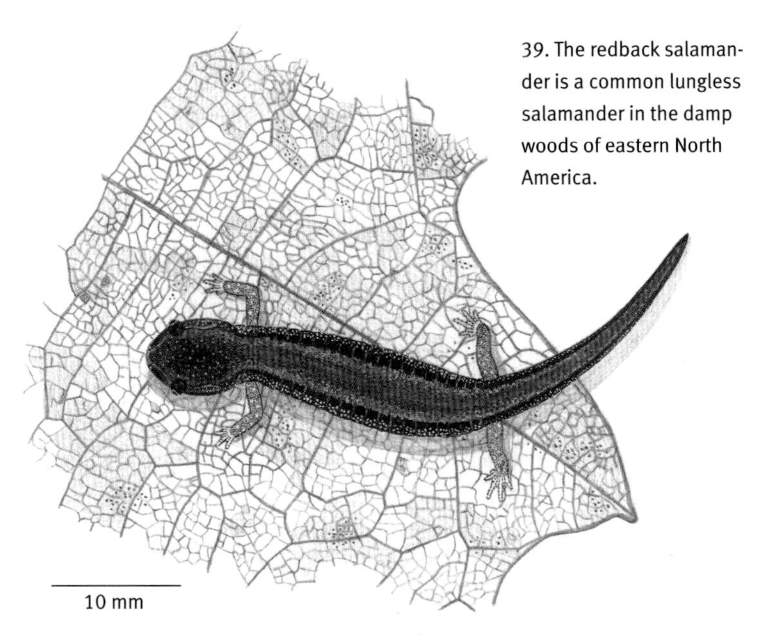

39. The redback salamander is a common lungless salamander in the damp woods of eastern North America.

40. Each toad has a small territory and a small burrow where it can be regularly found.

41. Spadefoot toads make rare appearances aboveground, usually only at night and during heavy rains. (Michael Jeffords)

42. Most skinks are swift runners and some are accomplished burrowers in sandy soils. (Michael Jeffords)

43. This legless lizard inhabits the dunes along the coast of California and Baja California. It can literally swim through sand. (Ken Lucas/Visuals Unlimited)

44. Ringneck snakes are tiny, harmless snakes that twist, coil, and expose their bright bellies when they are threatened. (Michael Jeffords)

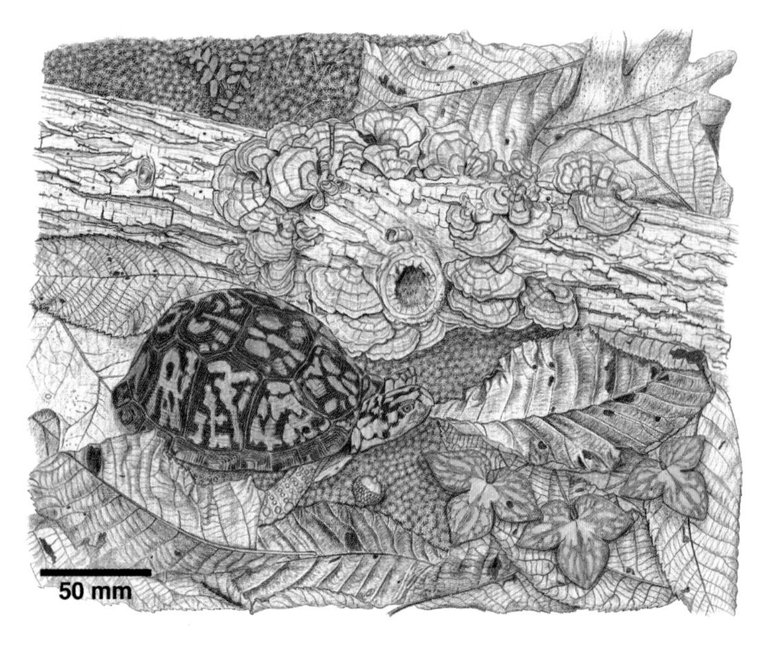

45. At the end of each day, a box turtle settles down in the leaf litter and mosses of its home territory. It builds a simple domelike chamber in which it nestles each night.

46. At sunset in the Arizona desert, a tortoise leaves its burrow and sets off for an evening of cactus grazing. (Susan Post and Michael Jeffords)

47. A robin listens for earthworms that are pulling fallen leaves into their burrows.

48. Magellanic penguins nest in burrows on the treeless islands of the Southern Hemisphere. (Gerald and Buff Corsi/ Visuals Unlimited)

49. Burrowing owls usually move into abandoned mammal burrows but can dig their own burrows when necessary. (Adam Jones/Visuals Unlimited)

50. Bank swallows nest in colonies of 10 to 100 pairs on steep sand or gravel banks. Each pair digs a horizontal burrow two to three feet (60 to 90 centimeters) deep. (Steve Maslowski/Visuals Unlimited)

51. The badger's long digging claws account for its prowess as a burrower. (Michael Jeffords)

52. Throughout the months of winter, thirteen-lined ground squirrels hibernate in their burrows. After arousing from their winter's sleep, they mate; and several weeks later their young venture forth from the burrows for the first time. (Michael Jeffords)

53. A chipmunk in Yellowstone National Park spends a good portion of its short summer gathering seeds and nuts to store in its underground den. (Michael Jeffords)

54. The star-nosed mole inhabits the damp, boggy soils of eastern North America. (Gary Meszaros/Visuals Unlimited)

55. Naked mole rats live in colonies and share the labor of digging their tunnels. (M. J. O'Riain and J. Jarvis/Visuals Unlimited)

56. Hedgehogs are found throughout the Old World except on Madagascar and Australia. (Joe McDonald/ Visuals Unlimited)

57. The jumping mouse probably spends more than half its life underground, hibernating in its nest of dry grass or moss during the colder months. (Inga Spence/Visuals Unlimited)

58. Conventional tillage destroys habitats for soil creatures, compacts the soil, and exposes it to wind and rain erosion. (Michael Jeffords)

59. By leaving some soil untilled, the farms in this photograph practice conservation tillage. The farmers minimize disruption of soil structure and soil habitats by minimizing their use of farm equipment. In combination with fence rows and windbreaks, land that is infrequently tilled attracts not only a diverse population of soil creatures but also birds and insects that help keep in check the populations of crop pests. (Michael Jeffords)

60. Without its mantle of vegetation and animal life, a soil loses its spongy structure and the nutrients that it held. (Ron Spomer/Visuals Unlimited)

61. The nutrients that the crops and livestock of a farm take from the soil can be best replenished by adding organic matter in the form of compost, animal manure, and green manures. (Michael Jeffords)

62. Acid rain that has fallen on this forest in eastern North America has weakened or killed many of its trees. (Adam Jones/Visuals Unlimited)

63. White crusts form on surfaces of soils that are high in sodium. By disrupting the structure of soil, the presence of sodium impedes the penetration of water and the growth of roots. (Fritz Polking/Visuals Unlimited)

64. Overgrazing compacts soil and removes vegetation that adds organic matter to soil and protects it from erosion. (Rudolf Arndt/Visuals Unlimited)

65. Clear-cutting of mountain forests has exposed soil to extensive erosion and nutrient loss on Madagascar. (Walt Anderson/Visuals Unlimited)

66. The act of composting enhances the well-being of both the soil and the composter. (Michael Jeffords)

67. Steam rises from a compost pile as microbes decompose organic matter and generate enough heat to raise the temperature as high as 165°F. (John Sohlden/Visuals Unlimited)

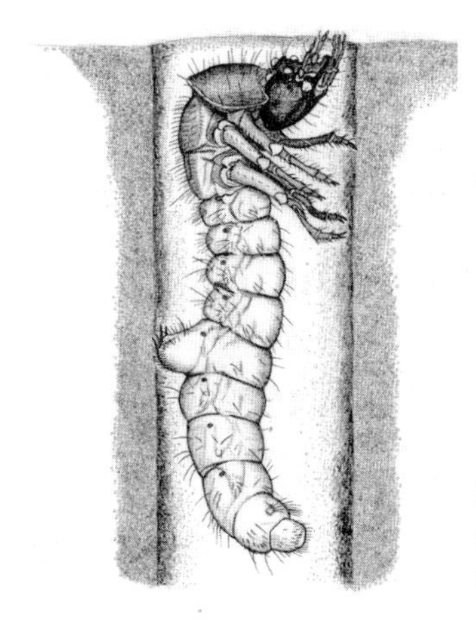

103. At the entrance to its vertical burrow, a tiger beetle larva waits to ambush a passing insect.

Rather than searching for its food, the larva lies in its burrow waiting for prey to pass by. By placing its jaws right at the entrance to the burrow, it seals the opening with its broad head and part of its thorax. Since the larva covers the exposed portions of its head and thorax with soil particles from the edge of the burrow, it becomes almost invisible. By maintaining a firm grip on the vertical wall of its burrow with the sharp hooks on its abdomen, it is ready to tackle most insect passersby and to subdue even insects larger than itself (fig. 103). The tiger beetle larva eventually drags its prey deep into the burrow and devours it at the bottom of the pit.

While tiger beetle larvae use subterfuge and stealth to ambush their prey, adult tiger beetles use speed in the air and on the ground to chase down other insects. The glistening colors of these flashy beetles make them easy to spot; but their long legs, sharp eyes, and strong flight muscles make them hard to catch (plate 26). They are constantly wary and rarely turn their backs on pursuers. The life of a tiger beetle is in many ways like the tale of the ugly duckling: from a grotesque larva that lives in a dark burrow comes a lovely and graceful beetle found in sunny clearings.

13. Short-winged Mold Beetles

Short-winged mold beetles or pselaphine (*pselaph* = grope around) beetles may be found among the molds of the soil, but they are also on the prowl for other soil invertebrates: springtails, mites, sym-

Phylum Arthropoda	**Place in food web:** predators, fungivores
Class Insecta	**Impact on gardens:** allies
Order Coleoptera	**Size:** 0.5–5.5 mm in length
Family Staphylinidae	**Estimated number of species:** 8,300
Subfamily Pselaphinae	

phylans, diplurans, and even small worms. These tiny beetles have beautiful beady antennae that expand at their tips and give them a very appealing and distinct character. The pselaphines, as their name implies, use these sensitive antennae to full advantage as they grope and wend their way through the labyrinthine passageways of the soil (fig. 104).

Like other members of the rove beetle family, the pselaphine beetles have short wing covers, or elytra, that only cover a portion of their abdominal segments. Normally elytra restrain the movement of a beetle's abdomen, so with such short elytra, these beetles are exceptionally limber and are free to move their long abdomens in just about any direction as the need arises. The pselaphine beetles that live the deepest in the soil are those with the shortest legs, the shortest elytra, and the fewest abdominal segments covered by the elytra.

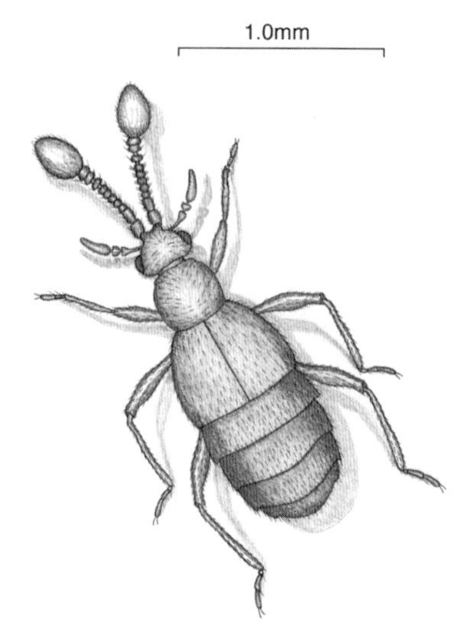

104. In addition to living under logs, rocks, and decaying leaves, short-winged mold beetles dwell in caves, burrows of mammals, and ant or termite colonies.

14. Featherwing Beetles

Phylum Arthropoda	**Place in food web:** fungivores
Class Insecta	**Impact on gardens:** allies
Order Coleoptera	**Size:** 0.4–1.5 mm in length
Family Ptiliidae	**Estimated number of species:** 400

The featherwing, or ptiliid (*ptilon* = feather), beetles include the smallest known beetles. Most members of the family measure less than a millimeter in length. Their wings, however, are usually more than twice this length and look more like tiny bird feathers or the fine tufts on dandelion seeds than insect wings. A few fungal spores are just about a mouthful for one of these beetles (figs. 105–6).

Even though a mother beetle lays only a single large egg at a time, she puts a great deal of energy into this egg because it can be as large as half her own length. What emerges from this egg is a miniature, scaled-down version of a rove beetle larva but without pigment, eyes, and sickle-shaped jaws. Some featherwing beetles lay normal, fertilized eggs; but many mother featherwings are virgins and lay only unfertilized eggs that produce only daughters. The progression from egg to adult for these little beetles takes only about a month, so even though these beetles are rarely noticed, they are still very abundant on the forest floor.

15. Sap Beetles

Phylum Arthropoda	**Impact on gardens:** allies/adversaries
Class Insecta	**Size:** mature larvae 1–15 mm in length
Order Coleoptera	**Estimated number of species:** 3,000
Family Nitidulidae	
Place in food web: decomposers, detritivores, scavengers, fungivores	

The sap beetles or nitidulids (*nitid* = shining; *-uli* = small) love saps of all sources, at all stages of fermentation and decay: sap from rotting fruit, sap flowing from trees, sap from fungi above- or belowground. In the soil, fermentation and decay are so pervasive that adult and

105. This featherwing beetle is shown folding its feathery hind wings under its hard fore wings.

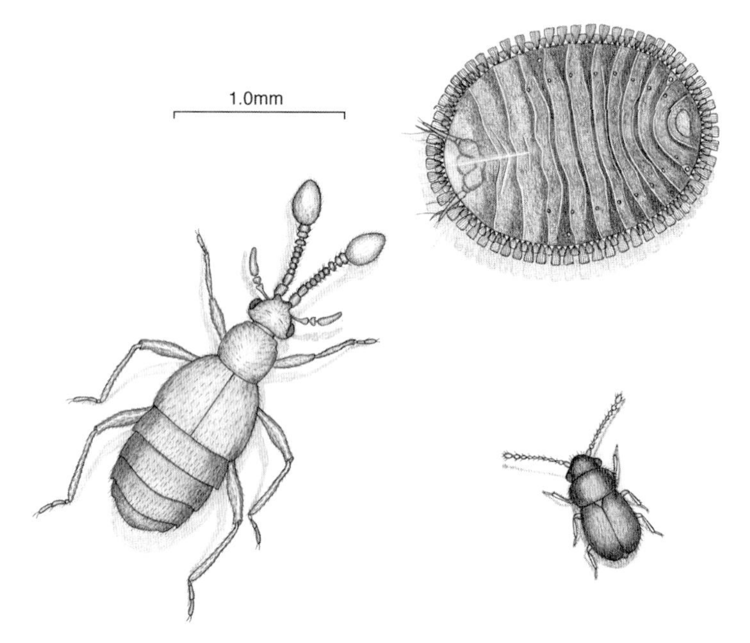

106. Featherwing beetles are dwarfed by the other beetles that also inhabit the decomposing leaves on the forest floor

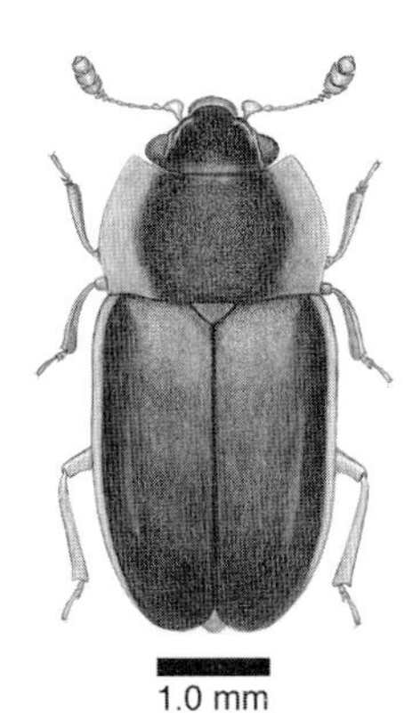

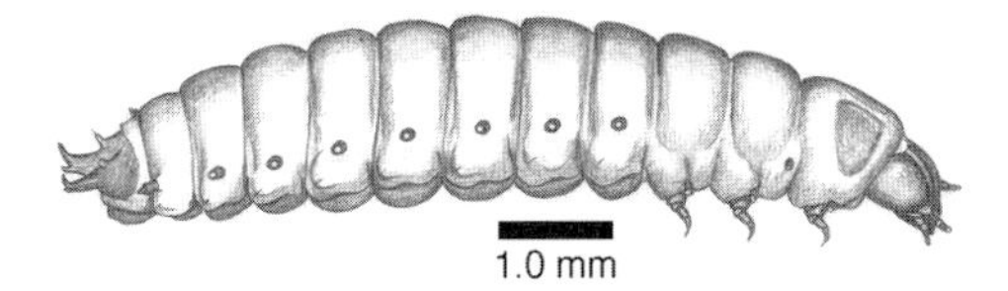

107–8. Most species of sap beetles live aboveground on green plants, but a few species of sap beetles (107, left) and their larvae (108, above) dwell in the soil among fungi and rotting plants.

larval sap beetles can feed on a variety of saps with little possibility of ever depleting their food source (figs. 107–8).

Not all nitidulids lead such placid lives. Along the foraging trails of certain ants live some notorious nitidulid beetles that act like highwaymen. During the hours of daylight these beetles lie in the leaf litter, but at nightfall they patrol the ant trails looking for some unsuspecting ant that might part with the food it is carrying back to the nest. By tapping its antennae on the ant's mouthparts and using the signals that ants use among themselves, these highwaymen induce the ant to regurgitate the food it is carrying. Often the ant realizes she has been misled by the beetle into believing that she is sharing her food with another ant. In retaliation for this deception, she attempts to lay her mandibles on the beetle; but by the time she does, the beetle has flattened itself on the ground, pulling all legs and antennae under its body. It grips the ground with strong bristles on its legs so the ant is neither able to attack it from above nor to flip over the beetle and attack it from below. As soon as the outwitted ant moves on, the beetle waits for the next ant it will deceive and pillage.

16. Antlike Stone Beetles

Although antlike stone beetles can easily be mistaken for ants at first glance, a closer look shows that their antennae are too straight to be ant antennae and their bodies are too hairy to be ant bodies. Like nitidulid beetles, many of these scydmaenid (*scydmaen* = angry) beetles have found ways to deceive and exploit other soil insects, living as scavengers in ant or termite colonies. Although the ants and termites probably can tell the difference between these beetles and the mem-

Phylum Arthropoda
Class Insecta
Order Coleoptera
Family Scydmaenidae
Place in food web: scavengers, predators
Impact on gardens: absent
Size: 0.6–2.5 mm in length
Estimated number of species: 3,570

bers of their own colony, they somehow tolerate these intruders and treat them with indifference. Antlike stone beetles that roam the leaf litter and hollow trees beyond the confines of ant and termite colonies survive both as adults and larvae on a diet rich in oribatid mites. Breaking through the hard shells of oribatid mites undoubtedly demands some sharp and hard mandibles (figs. 109–10).

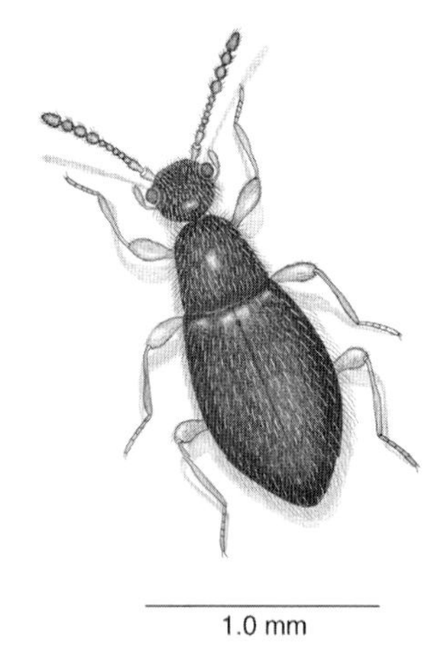

109. Antlike stone beetles live in the leaf litter, under bark, or as scavengers in ant and termite colonies.

110. Larvae of antlike stone beetles are small predators with prominent clubbed antennae.

17. Minute Fungus Beetles

Phylum Arthropoda
Class Insecta
Order Coleoptera
Family Corylophidae
Place in food web: fungivores
Impact on gardens: absent
Size: mature larvae 1–3 mm in length
Estimated number of species: 400

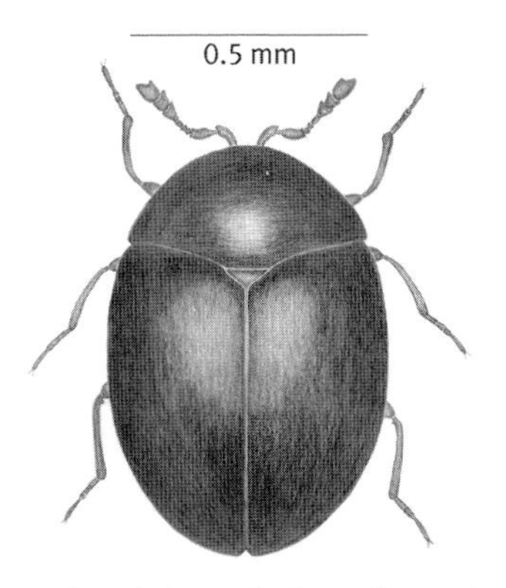

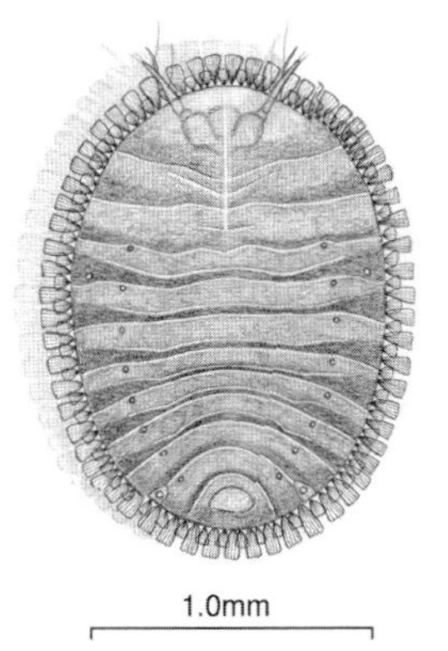

111–12. Populations of minute fungus beetles (111, left) and their larvae (112, right) live in the flat world between layers of decaying leaves. There they graze on fungi and spores that coat the leaf surfaces.

The most unforgettable beetle larvae of the leaf litter are probably those of minute fungus beetles, or corylophids (*cory* = helmet; *lophi* = crown). They are flat, broad, and shaped like a softshell turtle. The scientific term for this body shape is cheloniform (*chelona* = turtle; *form* = shape). A series of flat, ornate scales are arranged around the edge of the larva's body, similar to the colorful scales that cover butterfly and moth wings. On the back of the larva are openings of glands that probably provide protection from predators that prowl between the fallen leaves. It is easy to imagine one of these tiny saucer-shaped larvae roaming the narrow spaces between layers of fallen leaves on the forest floor. Soil fungi wend their way between the fallen leaves, where their spores and hyphae offer a feast for these little larvae.

In the adult beetles, the first segment of the thorax, known as the pronotum, extends forward over the head, hiding the eyes, the mouthparts, and the bases of the antennae. This protruding portion of the thorax was apparently construed as a crown or helmet by the person who chose the family name Corylophidae for these beetles. This "helmet" for the beetle's head probably helps the beetle plow its way through layers of leaves on the forest floor. These rotund beetles, along with the featherwing beetles, have long hairs adorning the margins of their hindwings and are some of the smallest of beetles. They are among the many adult beetles that, along with their larvae, are found in the company of molds and fungi (figs. 111–12).

18. Ptilodactylid Beetles

Phylum Arthropoda
Class Insecta
Order Coleoptera
Family Ptilodactylidae
Place in food web: decomposers, detritivores, scavengers

Impact on gardens: allies
Size: adults 4–10 mm in length
mature larvae 3–25 mm in length
Estimated number of species: 450

In early spring, bristly larvae of ptilodactylid beetles (or winged-toe beetles) appear in forest soil and litter, under the carpet of wildflowers that cover the forest floor. These larvae help convert the abundant leaf litter into the humus that nourishes the rich flora of the forest. Judging from the frequency with which they are encountered during the spring, these larvae must contribute a good share to the enrichment of the forest soil. But the distinctive adult beetles are rarely seen until later, at evening lights in late June. At that time of year, they turn out to be frequent visitors. The beetles move about quickly, waving their long, comblike antennae if they are males; they wave equally long, but less handsome, sawtoothed antennae if they are females.

If one looks closely at their feet, the origin of their family name

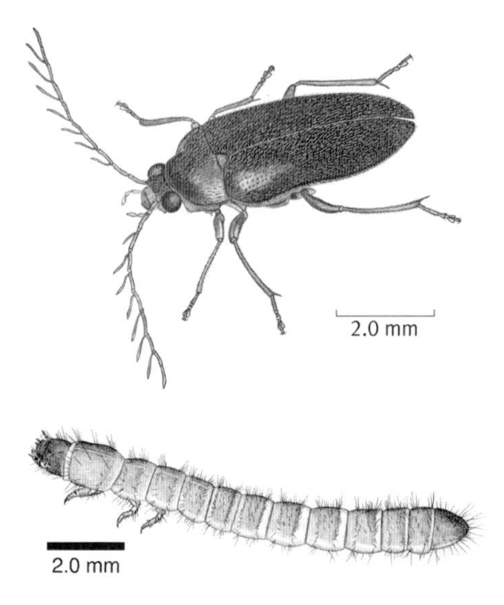

113–14. Ptilodactylid beetles are rarely seen, but down in the soil their larvae (114) are busy making important contributions to the recycling of leaves, roots, and wood.

becomes apparent. Like most scientific names, the long, five-syllable name for this family of beetles points out a distinguishing feature of these insects. Near the very end of each foot of a ptilodactylid (*ptilo* = wing; *dactyl* = toe), a large, winglike lobe extends down and back, almost to the tip of the claws. Many small beetles, like other small creatures, look nondescript and featureless until one is treated to close-up views of them. Through the lenses of microscopes features appear that surprise even the most seasoned observer of the living world (figs. 113–14).

19. Glowworms, Fireflies, and Lightningbugs

Phylum Arthropoda	**Impact on gardens:** allies
Class Insecta	**Size:** adults 4.5–20 mm in length
Order Coleoptera	mature larvae 17–50 mm in length
Family Lampyridae	**Estimated number of species:** 1,900
Place in food web: predators	

Every June and July, those who live in the eastern half of North America are treated to the sight of thousands of flashing fireflies lighting the evening skies. Between the Great Plains and the Atlantic Coast about 150 species of fireflies, or lampyrids (*lampyri* = glowworm), can be found. In the rest of the world more than ten times this number are found, most in the tropics. In North America, the biggest contributors to the midsummer displays of lights as well as the most familiar of the fireflies or lightningbugs are the 28 species of *Photinus* and the 20 species of *Photuris.*

Even though not all adult fireflies flash, all glowworms—as firefly larvae are usually known—turn out to be flashers. Many fireflies even start out in the soil as glowing, luminescent eggs. If we looked at the soil more often and more carefully on dark, moonless nights in summer and autumn, we might see that the ground at times can be thickly covered with their lamps; several hundred glowworms can often occupy a square meter of soil. As fast-moving predators, these larvae seize their prey with sickle-shaped jaws and inject both a neurotoxin that paralyzes their victims as well as digestive juices that help liquify muscles, nerves, and other tissues that can then be sucked back along a groove in each jaw. Larvae of the genus *Photuris* apparently

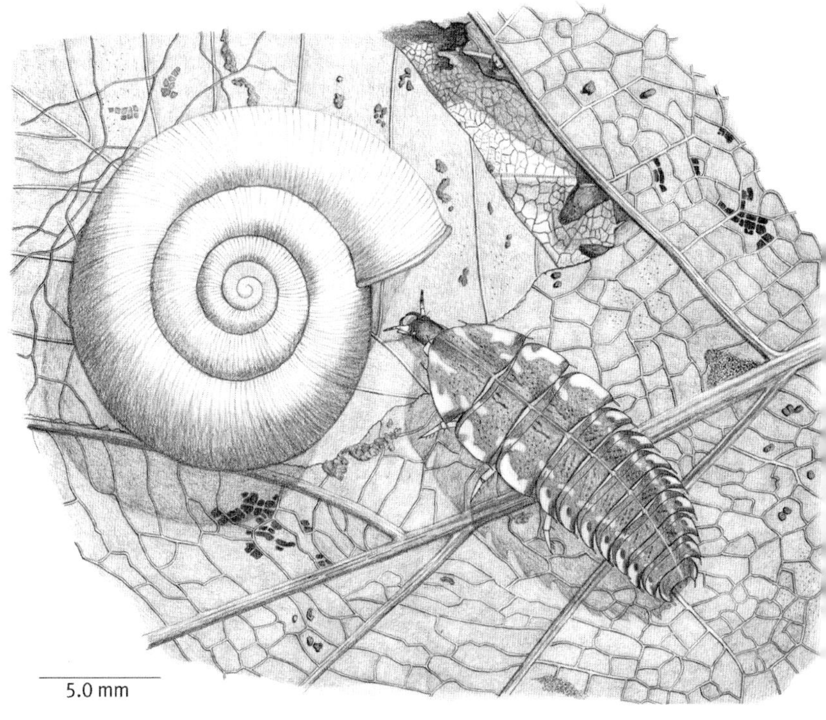

115. Larvae (glowworms) of the firefly genus *Photuris* are predators of snails and slugs.

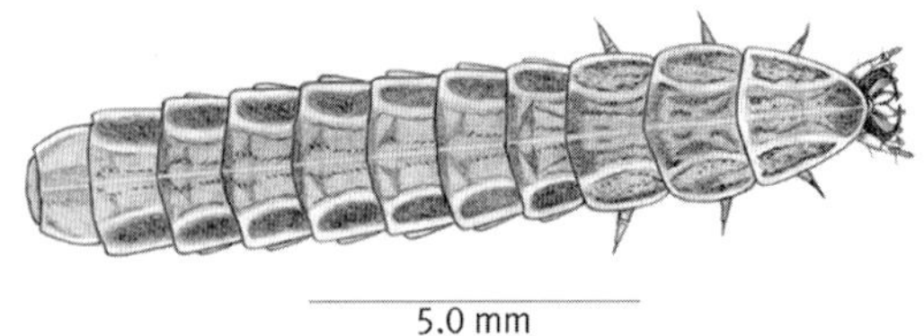

116. Larvae of the firefly genus *Photinus* are predators of earthworms.

prefer feeding on snails and slugs (fig. 115), while the more slender larvae of the genus *Photinus* search the narrow underground passageways frequented by earthworms (fig. 116). The European glowworm *Lampyris* has been observed tracking down snails and slugs by following their slime trails through the leaf litter. At the approach of winter, the larvae burrow into the soil; in the spring they transform to pupae and adults.

The adults soon spread their wings and set about the business of courting. The males cruise over fields and clearings, each species flashing its own distinctive code, which is translated by the females as "Here I am, and I'm available"; and the females fly off to a plant or spot on the ground where they can pick and choose their favorite flashers (plate 27). The flashing patterns for the different species are so distinctive that most of these beetles can be recognized from the number, duration, and intervals between these flashes. Although adults flash as part of their courtship ritual, eggs and larvae light up to warn toads and other possible predators of the soil surface that glowworms not only do not taste very good but they are also downright toxic.

Sadly, many areas are losing the magic that fills the air on summer evenings, when hundreds, often thousands, of fireflies perform. Fireflies are becoming rarer wherever soil habitats that support firefly larvae and the earthworms, snails, and slugs on which they feed are paved over, plowed under, and eliminated by human encroachment on the few remaining wild places.

20. Soldier Beetles

Phylum Arthropoda
Class Insecta
Order Coleoptera
Family Cantharidae
Place in food web: predators

Impact on gardens: allies
Size: adults 1–15 mm in length
mature larvae 5–20 mm in length
Estimated number of species: 5,100

Although these beetles neither flash nor glow, they still share a number of other attributes with their close relatives the fireflies and glowworms. Most soldier beetle larvae run about in the soil and leaf litter, dispatching other soil creatures with their large, sickle-shaped jaws. Like the larvae of fireflies, these larvae have grooves on the surfaces of their mandibles that drain the body juices of their victims. Their life cycle also closely parallels that of fireflies. After a summer of prowling and hunting, the mature larvae retire to sheltered spots in the soil and leaf litter, where they stay through the winter. Lengthening days and rising temperatures of spring signal that the time has

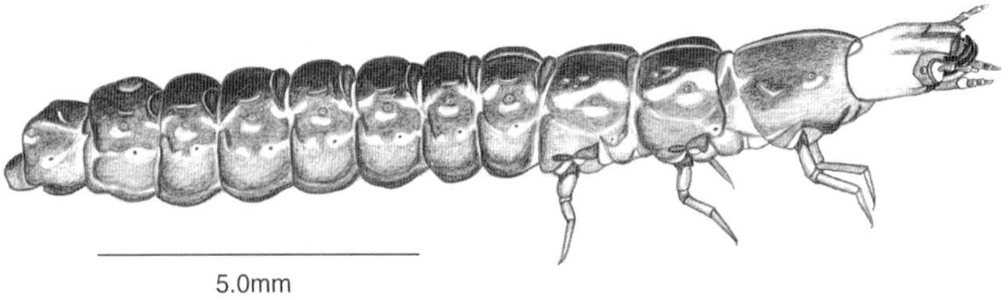

5.0mm

117. The dark, velvety larvae of soldier beetles are fast-moving predators of the soil.

come to pupate and leave the soil. While fireflies court in the evening skies of early summer, soldier beetle romances begin on sunny days among the flowers.

A number of features set soldier beetle larvae apart from other beetle larvae. Larval soldier beetles are covered with a dense coat of short, fine hairs that makes them look and feel as though they are attired in velvet. Deep shades of brown, purple, burgundy, and black are the colors of the coats of most larvae, but those destined to become aphid eaters as adults have pink velvety coats (fig. 117). Soldier beetle larvae repulse most potential predators with the noxious chemicals produced in stink glands on their backs. Their chemical repellents and their velvet coats give these larvae an unmistakable look and a distinctive smell.

Adult soldier beetles have soft, leathery bodies and are almost always decked out with a combination of orange or yellow and black or brown pigments. They often congregate in large numbers on flowers, where they share nectar and pollen with bees, flies, and butterflies (plate 28). Some of the darker soldier beetles gather on plants where aphids or scale insects (coccoids) are feeding and share their meals of aphids with ladybugs and lacewings. Once you start looking for these beetles, you will discover that they are quite common and colorful; they just are not as flashy as their relatives the fireflies.

21. Dung Beetles

Certain beetles dine solely on dung and share their stinky meals with fungi, nematodes, and flies. Transforming dung to humus is a faster process than converting dry grass, sawdust, or an oak leaf to humus. Organic matter in dung is not only more finely divided than an oak leaf but it is also much richer in nitrogen. There are no shortage of

Phylum Arthropoda
Class Insecta
Order Coleoptera
Family Scarabaeidae
Place in food web: decomposers, detritivores, scavengers, coprophages, diggers
Impact on gardens: allies
Size: 2–30 mm in length
Estimated number of species: 7,500

insects that compete with one another for coveted dung piles and that are instrumental in converting them to humus. But they can be very particular about whose dung they recycle (fig. 118).

All the continents of the world but Australia originally had native populations of large grazing animals like cattle, horses, elephants, and buffalo; each of these animals consumes large quantities of vegetation and can drop enough dung pads in one year to cover a tenth of an acre of soil. Fortunately, these grazing animals share their habitats with beetles that have found a use for the dung pads and ensure that the soil surface does not become smothered beneath a layer of dung. No one really thought too much about the usefulness of these dung beetles, and their achievements were mostly taken for granted until cattle, horses, and sheep were introduced to Australia and dung

118. All animals contribute dung to the soil, but dung of some animals enriches the soil more than that of others.

pads quickly began to overrun pastures. The native Australian dung beetles had not evolved to handle the dung pads of the large livestock imported from Europe and showed no inclination to do so. Only after dung beetles were imported from Africa and southern Europe was the Australian dung problem solved.

Kangaroos had been the largest mammals to graze the pastures of Australia, and they left dry dung pellets no larger than golf balls. Even though some dung beetles have relatively undiscriminating tastes, many of them are very finicky about their choice of dung pads. Only those pads with particular textures, vegetable composition, water content, or specific dimensions are acceptable. For the Australian dung beetles, kangaroo dung was an acceptable meal but cow dung was not.

The dung beetles of Africa are accustomed to dung pads of every imaginable size, texture, and composition: from the large imposing pads of elephants and rhinos to the small pellets of antelopes and gazelles. Even before dung hits the ground, some dung beetles have picked up its scent and rush to claim these highly prized resources. Elephant dung alone is the preferred habitat for 150 species of dung beetles. A count of more than 7,000 of these beetles on a pad of fresh elephant dung hints at just how popular dung pads can be. Adult beetles feed on the more nutritious juices of the dung, and they move solid dung to underground chambers. With help from the bacteria in their guts, the larvae are able to digest even the cellulose fibers that make up so much of a herbivore's dung. Also by eating their own droppings, these larvae enrich their diet with all the bacteria that pass along with the droppings.

Some of the chambers that adults provision for their larvae are elaborate and extensive; one species of beetle may leave up to forty dung balls in its underground chambers. Others excavate only shallow pits in the soil and dexterously maneuver dung balls that are often larger than themselves into these pits. Dung beetles clearly take great care in providing a dung ball for each of their larvae. By constructing their nests underground, larvae are protected from predators, parasites, competitors, and often from bad weather.

As part of their ancient routine of excavating chambers to accommodate their dung balls and their larvae, all dung beetles participate in mixing nutrients from dung with minerals of the soil. This activity increases the permeability of soil to air and water, increases its

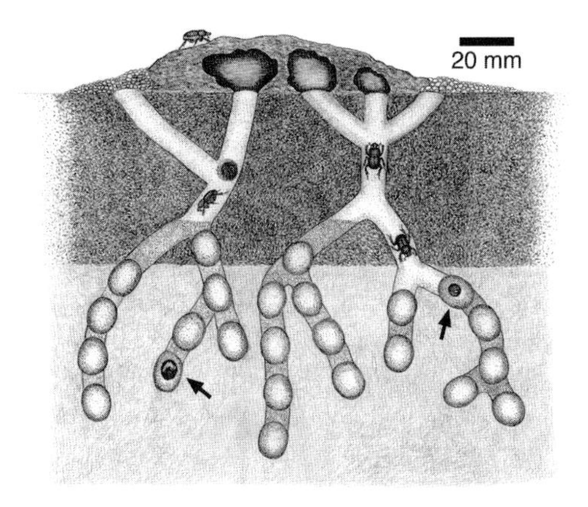

119. As beetles position their dung balls beneath a dung pad, dung goes down and subsoil goes up. The arrow to the left points to a dung ball cut open to reveal a beetle larva; the arrow to the right points to a dung ball cut open to show a dung beetle egg.

organic content, and improves the soil structure. In the process of doing all this, dung beetles can bury up to a ton of dung on a hectare of soil (fig. 119, plate 29).

An elephant or a cow may be efficient about obtaining energy from the food it eats, but it doesn't quite extract all the energy. All animal droppings retain some food and energy value. Someone once calculated that in the course of a year, one cow leaves enough food and energy in its droppings to support an insect population equivalent to at least 20 percent of the cow's weight as well as millions of bacteria and thousands of nematodes, protozoa, and fungi. The Plains Indians of North America were well aware that buffalo dung was a rich source of energy. They cooked and warmed themselves by fires fueled with dried buffalo dung. The dung of herbivores, or plant-eating animals, is high in energy value. The food consumed by carnivores, or meat eaters, however, contains little energy value after it is converted to droppings. Carnivore dung neither makes a good fuel for a fire nor does its poor energy value attract many hungry insects.

Other insects are also fond of dung, and struggles often arise over who lays claim to this coveted resource. Fly maggots and dung beetles are the two main contenders for every new pile of dung. Dung beetles happen to have a number of allies in their clashes with flies and their maggots. Among their larger allies are shiny black hister beetles and sleek, sinuous rove beetles that spend most or all of their lives in dung pads, where they prey on fly maggots and fly pupae. Like dung beetles, hister beetles and rove beetles usually carry mites as tiny allies on their bellies and under their wing covers, or elytra. When the beetles land on the dung pads, the mites leap off and scurry about in search of fly eggs or newly hatched maggots. Each beetle can carry up to thirty mites; and if a beetle somehow loses its mites, it simply cannot compete as well with the maggots for its fair share of the dung pad. These mites can significantly reduce the population of maggots before maggots are able to take over the dung pad. After beetles and mites satisfy their appetites in the dung and before the last beetles leave the dung pad, mites hop aboard the beetles and set off for new destinations.

22. Carrion Beetles, Burying Beetles, and Hister Beetles

Phylum Arthropoda
Class Insecta
Order Coleoptera

Carrion Beetles and Burying Beetles
Family Silphidae
Place in food web: scavengers, detritivores, decomposers, predators
Impact on gardens: allies
Size: 12–40 mm in length
Estimated number of species: 215

Hister Beetles
Family Histeridae
Place in food web: predators
Impact on gardens: allies
Size: 0.5–20 mm in length
Estimated number of species: 3,700

A dining experience that sounds as unappetizing as a dinner of dung is a meal of decaying carcasses. Carrion beetles are the vultures of the insect world and see to it that the corpses of animals are returned to the soil that nourished them during their lifetimes. Their clubbed antennae are finely tuned to the potent odors of dead animals—always animals with backbones. Fierce competition often develops among beetles for the limited number of dead animals that happen to be on the ground at any given time or in any given place. Battles between beetles are often waged for possession of these valuable resources (fig. 120).

One type of carrion beetle, the colorful sexton beetle, or burying beetle, is named after the church official in charge of grave digging. These beetles are appropriately decked out mostly in black, but they also have splashes of red or orange to relieve their somber attire. Male and female beetles work as a team to bury their find as quickly as possible before fly maggots or other beetles can establish a foothold on the dead body. Small birds and mammals are obviously preferred

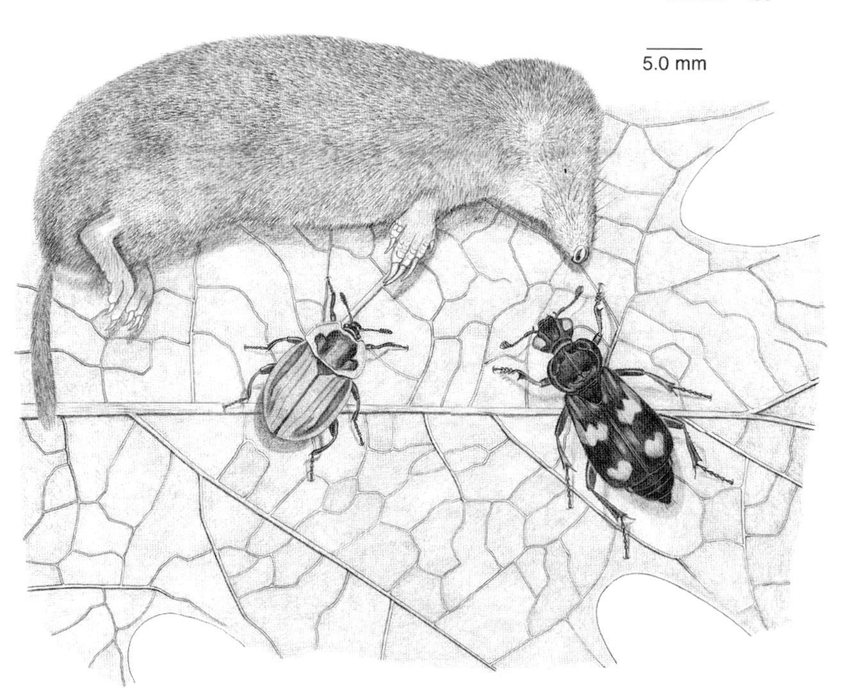

120. A dead shrew is a resource that is much coveted by carrion beetles. *Silpha* approaches from the left and a sexton beetle approaches from the right.

for logistical reasons; it is far easier for these inch-long beetles to maneuver a mouse than a cow. The generic name of the sexton beetle is *Nicrophorus* (*nicro* = dead; *phorus* = carry). Those fly eggs and fly maggots that hatch on the carcass that the beetles have claimed are soon dispatched by the beetles themselves or by the mites that travel with the beetles.

After a few hours of digging, shifting, and molding, the dead body is reduced to a round, unrecognizable ball buried several inches belowground. The mother beetle moves about this underground chamber, laying about 20 eggs along its walls. Next she begins chewing a little hole at the top of the ball of flesh that will serve as a nest for her larvae when they hatch in a few days. For the first few hours after the larvae have moved into the nest, the mother beetle feeds her brood just as a mother bird would feed her nestlings. After their second molt, however, the larvae are on their own and eat their way deeper into the ball of decaying flesh until they are ready for metamorphosis.

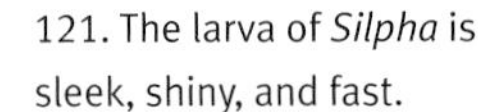
121. The larva of *Silpha* is sleek, shiny, and fast.

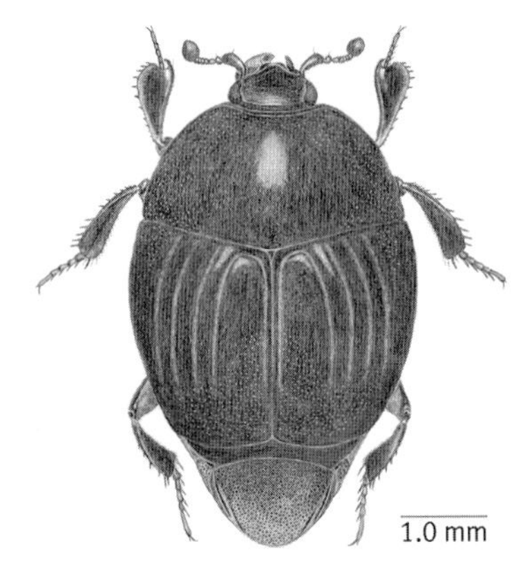

122. Many hister beetles have the luster of burnished ebony.

123. The larva of a hister beetle has imposing jaws but legs that are disproportionately small and skinny for its size.

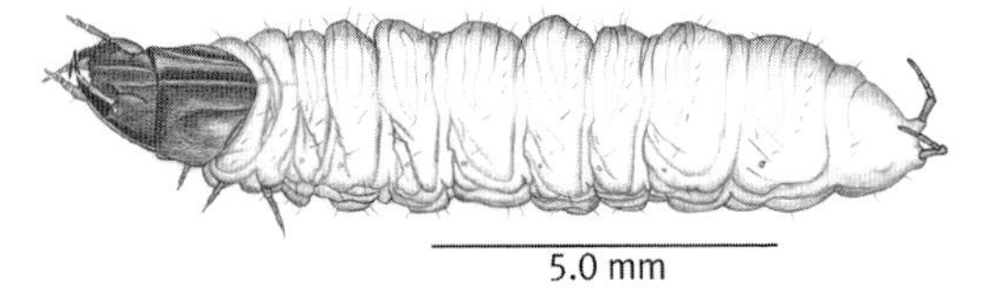

The other carrion beetle known as *Silpha* is nowhere near as sophisticated about its larval rearing. *Silpha* does not bury the corpses that its antennae lead it to nor is it particularly concerned about the size of the corpse. A dead horse or even a large, well-rotted mushroom will suffice as well as a dead shrew. *Silpha* and its larvae are not adverse to sharing their rotten find with maggots and other beetles, for they often add these insects to their meal (fig. 121).

Jet-black, oval, and shiny, many hister beetles keep company with the carrion beetles (figs. 122–23). Hister beetles survive primarily on a diet of maggots and any other smaller insects they come across. The lower segments of their legs are broad and spiny, clearly built for digging in dung, carrion, or fungi that are past their prime. A few hister beetles and their larvae are very flattened and live under bark of decaying trees; some are long and cylindrical, and live in the tunnels of wood-boring insects; still other members of this beetle family move into nests of ants or rodents. By reducing the populations of voracious maggots in carrion and dung, hister beetles are welcome allies for the dung beetles and carrion beetles with whom these maggots compete for very popular, but limited, resources. Most hister beetles also carry a crew of predatory mites wherever they go, and these mites begin devouring fly eggs and tiny maggots as soon as they alight on dung or a dead animal.

OTHER CARRION BEETLES

The agyrtid beetles are a group of creatures whose place in the family tree of beetles has changed a number of times in the last 150 years. Until recently they were a subfamily of the Silphidae. Now they are considered distinct enough to merit their own family, the Agyrtidae. This beetle family has only about 60 species; all of them are found at northern latitudes in the Northern Hemisphere with the exception of one far-flung species in New Zealand. Certain species of beetles in this family, such as the one shown here, seem to prefer decaying vegetable matter to decaying animals.

On the west side of North America, agyrtid beetles are some of the more conspicuous scavengers and detritivores of compost. The Farm at Chili Nervanos in Oregon practices organic farming and encourages the participation of as many creatures as possible in its composting process. Compost from Chili Nervanos has passed through the jaws of numerous agyrtid beetles as well as many other arthropods and annelids (figs. 124–25).

---

23. Wireworms and Click Beetles

Phylum Arthropoda
Class Insecta
Order Coleoptera
Family Elateridae
Place in food web: decomposers, detritivores, herbivores, predators

Impact on gardens: adversaries/allies
Size: adults 1.5–45 mm in length
mature larvae 10–60 mm in length
Estimated number of species: 9,000

Anyone who has worked the soil or looked inside a rotting log has most likely come across wireworms—long, sleek, and hard like an electrical wire. There may be many other insect larvae that dwell in

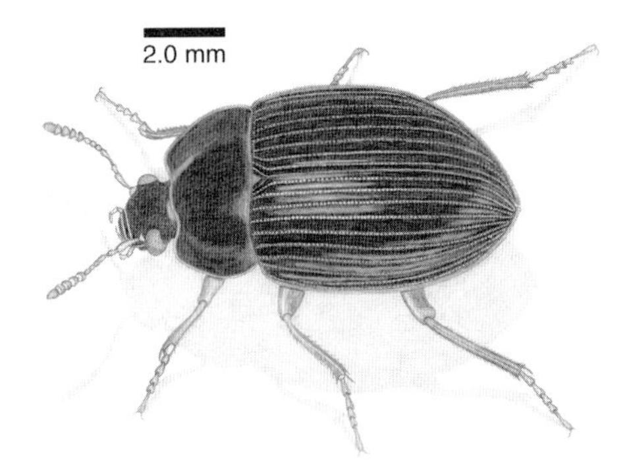

124. Classification of organisms can shift with time as demonstrated by agyrtid or primitive carrion beetles. These beetles that were once placed in the same family as carrion beetles (Silphidae) are now considered more closely related to another family of beetles known as the small carrion beetles (Leptodiridae).

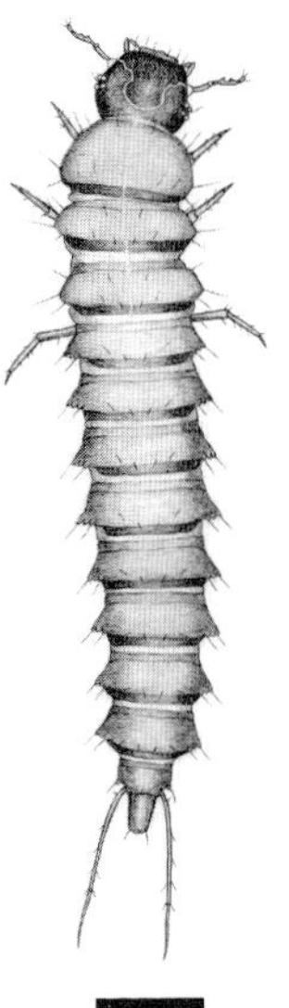

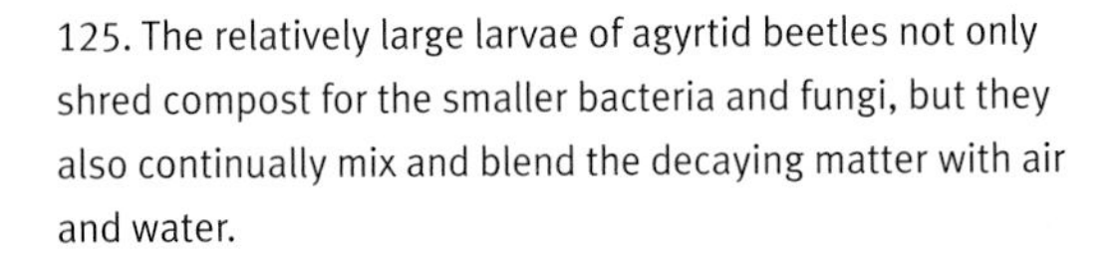

125. The relatively large larvae of agyrtid beetles not only shred compost for the smaller bacteria and fungi, but they also continually mix and blend the decaying matter with air and water.

the soil, but wireworms are among the most conspicuous. Each wireworm probably spends more of its life in the soil than any other insect with the exception of cicadas. Except for some of the smallest of the wireworms, most will spend at least three years and perhaps as many as seven years underground.

Wireworms are not only active but also dominant in a variety of subterranean roles: recycling plant debris, preying on other arthropods, and chewing on roots. Their sleek bodies slide easily through the soil; and although the larvae do not have eyes, the multitude of sensory bristles that covers their bodies from head to tail keeps them well informed about what is happening around them (fig. 126).

Wireworms eventually become the click beetles (see fig. 71), or elaterids (*elater* = hurler) of acrobatic fame, known for their ability to hurl themselves into upright positions if they ever land or are placed on their backs. An upside-down beetle cocks its head and part of its thorax back, forming an arch between them and the rest of its body. When the tension is suddenly released, the beetle's arched back

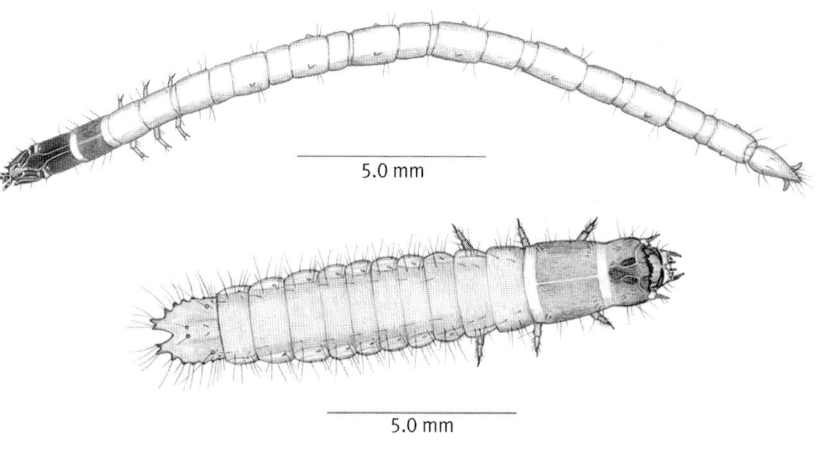

126. Wireworms live in soil, litter, or rotten wood. Some are long, thin, and sinuous but others are shorter, stouter, and armed with spiny tails, or urogomphi.

strikes the ground with such force that it springs into the air, most of the time landing upright on all six legs.

Wireworms can be a blessing as well as a curse in fields where crops are grown. Predatory wireworms prefer fields that are rich in plant debris, and they eliminate respectable numbers of insects, including other species of wireworms, that may be detrimental to the crops. The root-feeding wireworms, on the other hand, can inflict a significant amount of damage to crops once their population grows to more than a few hundred thousand larvae per acre. Root crops like potatoes and carrots seem to be especially sensitive to the ravages of these wireworms. By considering how the soil is prepared for planting each spring and prepared for winter each fall, soil conditions may be found that favor the populations of predatory wireworms. By encouraging these predators to eliminate enough of the root-feeding wireworms, pouring pesticides on the soil will become an unnecessary effort and expense.

24. Beetles of Rotten Logs

Dining on logs takes not only a good deal of chewing but also some help with the digestion of those tough wood fibers. Years before a dead tree falls to the ground, legions of bark beetles, wood-boring beetles, and wood wasps settle under the bark and in the heartwood as they begin preparing the tree for its eventual resting place in the

Phylum Arthropoda
Class Insecta
Order Coleoptera
Place in food web: decomposers, detritivores, fungivores
Impact on gardens: absent

Reticulated Beetles
Family Cupedidae
Size: 7–20 mm in length
Estimated number of species: 26

Telephone-pole Beetles
Family Micromalthidae
Size: 1–2.5 mm in length
Number of species: 1

False Click Beetles
Family Eucnemidae
Size: 3–35 mm in length
Estimated number of species: 1,300

Stag Beetles
Family Lucanidae
Size: 15–40 mm in length
Estimated number of species: 1,200

Patent-leather Beetles
Family Passalidae
Size: 30–40 mm in length
Estimated number of species: 500

soil. Microbes that live in the digestive tracts of wood-feeding insects have been recruited as the insects' best antidotes for indigestion. Larvae and adult insects depend on the bacteria, protozoa, and fungi of a rotting log and will slowly starve if they are fed logs in which these microbes have been killed. Larvae of many beetles and flies that feed on rotting wood also have specialized pouches in their intestines, where microbes reside that help digest some of the tough cellulose fibers that the larvae swallow.

In this respect these insects resemble large herbivores like cows and goats that have special chambers in their digestive tracts (rumens) teeming with bacteria, protozoa, and some fungi such as chytrids. These microbes help digest the large amounts of cellulose swallowed with each bite of grass or hay. The nitrogen content of the wood fibers swallowed by insects is probably so low that some of the bacteria of beetle and fly guts must convert or fix dinitrogen gas from the surrounding air to a form of nitrogen that insects can use to survive and grow (figs. 127–29).

Based on what we know about beetle genealogy, the first beetles that walked the earth about 240 million years ago were most similar in appearance and habits to the telephone-pole beetles and reticulated beetles that today live in moist and well-rotted logs. These are not only the most primitive living beetles, but they are also among the most unusual. By chance someone once discovered that male reticulated beetles find the odor of laundry bleach irresistible. The odor

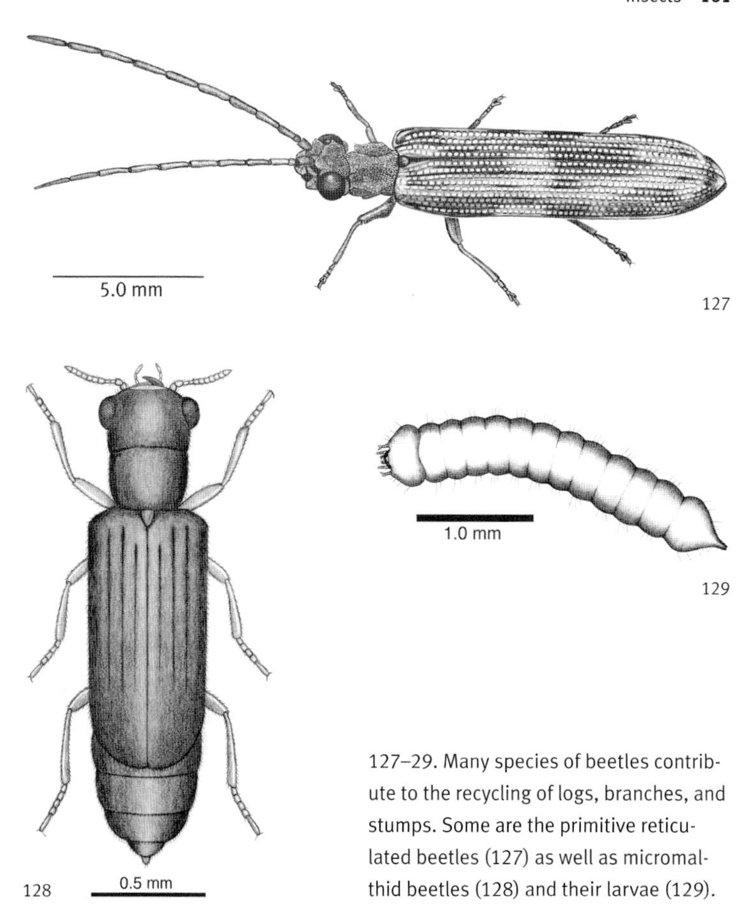

127–29. Many species of beetles contribute to the recycling of logs, branches, and stumps. Some are the primitive reticulated beetles (127) as well as micromalthid beetles (128) and their larvae (129).

is so similar to the attractant emitted by the female beetles that a few teaspoons of bleach are sufficiently alluring to entice these rarely seen beetles from their retreats in the deep woods.

Even fewer people have seen the tiny telephone-pole beetles. This unusual common name was chosen because these beetles have been found in old, rotting telephone poles; however, they are most often referred to as micromalthid beetles since they belong to the family Micromalthidae, which contains only one genus, *Micromalthus*. The largest specimens of this beetle measure only two or three millimeters in length. Although they have the reputation of being quite rare, the larvae can be found in large numbers in certain logs. They have a definite preference for logs in the oak family that are at just the right stage of decay. Once they have found just the right log, the larvae of micromalthid beetles begin multiplying as rapidly as possible to take full advantage of this ideal habitat before it disappears. To use the ephemeral resources of their log most efficiently, the larvae begin reproducing before they undergo metamorphosis and even before they mate.

These larvae are both paedogenic and parthenogenic, giving birth while they are still larvae (*paedo* = child; *genesis* = birth) and without mating (*parthenos* = virgin; *genesis* = birth). This policy clearly enables the beetles to share their resource with as many relatives as possible in as short a time as possible. No other beetle is known to use such a strategy. Within the beetle order Coleoptera, micromalthid beetles and reticulated beetles are considered unique enough to be placed in their very own suborder, appropriately named Archostemata (*arch* = ancient; *stema* = line).

The larvae of the various wood-boring beetles found in rotting logs often look very much alike. As their name implies, false darkling beetles of the family Melandryidae (*melan* = black; *dry* = tree) resemble darkling beetles of the family Tenebrionidae (*tenebrio* = lover of darkness). The adult beetles look alike, and the larvae do also. Larvae of darkling beetles are also known as false wireworms because they look like the wireworms that are larvae of click beetles. Wireworms that live in the soil generally chew on roots, but the wireworms of rotting logs chew mostly on the other beetle larvae that live there. As adults, another family of beetles (Eucnemidae) looks so similar to click beetles that they are referred to as false click beetles. Most wood-boring insects bore with the grain of the wood, but larvae of false click beetles bore across the grain; and they bore with jaws that curve out instead of in. Their eating habits must be quite unusual. Despite all the similarities among darkling beetles, false darkling beetles, click beetles, and false click beetles, the larvae of false click beetles are clearly different from all other beetle larvae (fig. 130).

The patent-leather beetles, or passalid beetles, like rotting logs that are well on their way to becoming humus. The adults hang around the logs and even present finely chewed wood pulp to their hungry, rapidly growing larvae. The squeaky noises that both larvae and adults produce represent a language that these beetles use among themselves. The noises or stridulations arise whenever a young or old beetle rubs together two different parts of its body. A larva uses

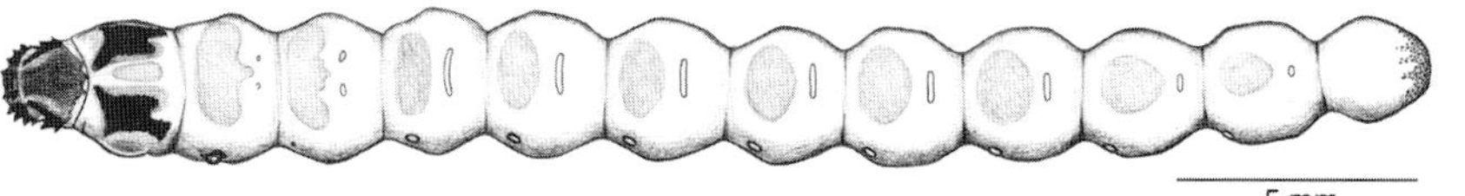

130. The larvae of false click beetles are strange in many ways. They are legless and eyeless, their antennae are absent or reduced to tiny stubs, and they have jaws that move out instead of in.

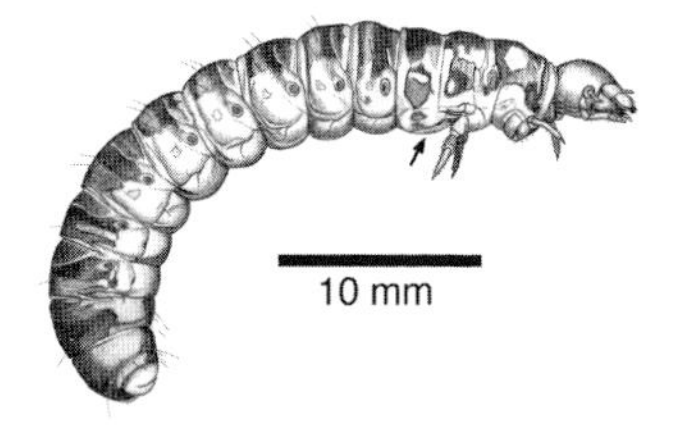

131. Passalid beetle larvae live in the same galleries within rotting logs as the adult beetles.

its short, stubby, and twisted third pair of legs, as indicated by the arrow in fig. 131, to strum rough patches of cuticle just above its second pair of legs. The adult beetle rubs a rough patch on the underside of its wings against another rough patch on the dorsal surface of its body (fig. 131, plate 30).

Larvae of the rather formidable-looking stag beetles, often inhabit the same well-decayed logs as their relatives, the passalid beetles. Stag beetle larvae look a great deal like their placid relatives, but the adult beetles can be as threatening as they look. Their large and pointed jaws can deliver a good pinch. Just as certain flashy beetles have been given the name lightningbugs, these beetles with king-size jaws are often called pinchingbugs. The large, lumbering larvae of pinchingbugs and patent-leather beetles devote many hours to wood chewing as they go about converting old, rotting logs to humus and extracting enough nourishment, with help from their gut microbes, to support their large bodies (plate 31).

25. Scarabs, Weevils, and Their Grubs

White grubs are sluggish, C-shaped larvae of beetles known as scarabs. These relatives of dung beetles include the well-known May beetles, June beetles, and Japanese beetles. Their appetites for roots of grasses, grains, legumes, trees, and shrubs combined with their abundance in the soil have earned these larvae reputations as devastating pests. To give you some idea just how abundant these white grubs

Phylum Arthropoda
Class Insecta
Order Coleoptera

Scarabs and White Grubs
Family Scarabaeidae
Place in food web: herbivores, diggers
Impact on gardens: adversaries
Size: 10–125 mm in length
Estimated number of species: 25,000

Weevils and Weevil Grubs
Family Curculionidae
Place in food web: herbivores, diggers
Impact on gardens: adversaries
Size: 2–30 mm in length
Estimated number of species: 50,000

can be, consider the following numbers. Larvae of Japanese beetles average about 175 per square meter in areas infested by the beetle, but one heavily infested golf course had a population of 1,500 larvae per square meter. For another species of white grub, the number of larvae counted in the same area was 4,500. One unfortunate corn plant was found to have 200 white grubs between half an inch and one inch (between 10 and 25 millimeters) in length feeding on its roots.

These larvae move up and down in the soil according to the season and the amount of precipitation. Some can descend as deep as five feet (150 centimeters) in cold or dry weather, yet at other times they feed only an inch (2.5 centimeters) below the soil surface. During their migrations through the soil, white grubs stir up the soil quite

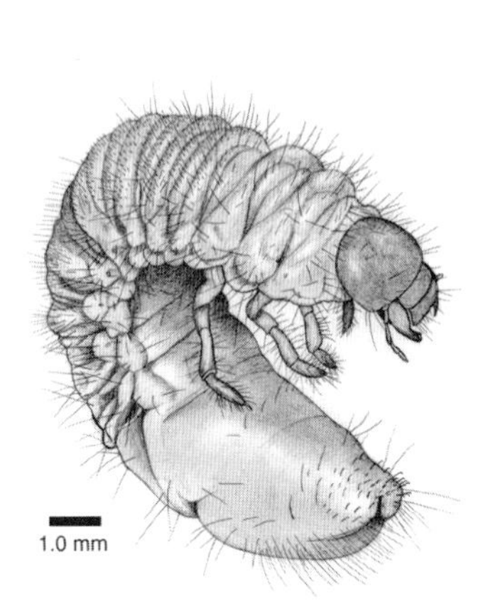

132. White grubs often spend several years underground.

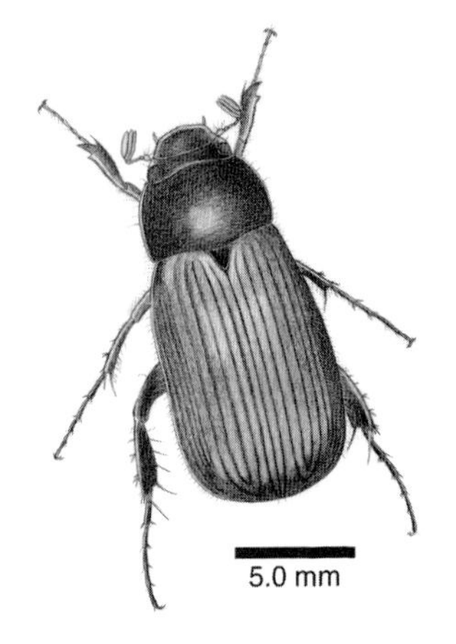

133. After feeding on roots as larvae, the adult beetles feed aboveground on leaves, flowers, and fruit.

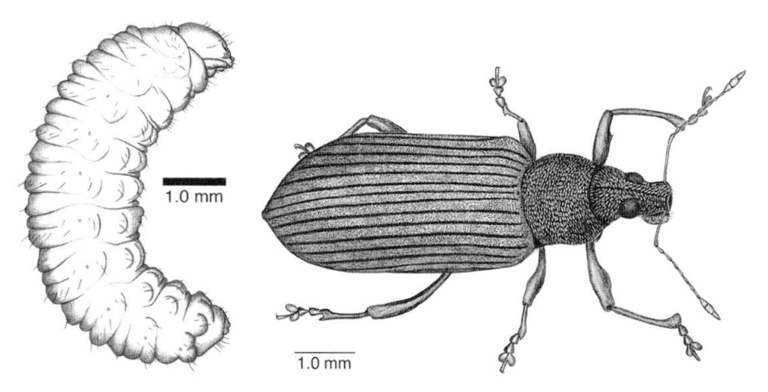

134–35. Weevil grubs (134, left) that feed on roots might be sluggish and sedentary, but they can inflict phenomenal damage on plants. As adult broad-nosed weevils (135, right), these beetles can strip many leaves from plants.

a bit. They might wreak havoc with the roots in the neighborhood, but they do improve soil structure and fertility for the roots that will take their place. Some white grubs, like those of Japanese beetles, mature in a single year; others feed on roots for two or three years. In colder climates some may feed underground for even four or five years (figs. 132–33).

The weevils, all of which are plant feeders, make up the largest family of beetles on the planet. All weevils have their jaws located at the ends of their prominent snouts, which are often referred to as "noses." Different weevils feed on different plant parts, and every plant part is eaten by some weevil somewhere. Some weevils use their snouts, which are longer than their bodies, for drilling into nuts, but broad-nosed weevils have shorter and broader snouts (figs. 134–35).

While adult broad-nosed weevils chew on plants aboveground, the legless larvae or grubs of these weevils nestle underground in cozy chambers within roots. Roots are eaten and occupied by grubs throughout the summer, fall, and winter. As the ground warms in spring, the full-grown larvae pupate and adults emerge in late spring. Adults and grubs usually share the same host plant, and together they can devastate entire fields of vegetables.

26. Variegated Mud-loving Beetles

Mud-loving beetles hang out in the muddy soil around the edges of ponds and streams. There are actually two families of closely re-

Phylum Arthropoda
Class Insecta
Order Coleoptera
Family Heteroceridae
Place in food web: diggers, decomposers, scavengers
Impact on gardens: absent
Size: 1–8 mm in length
Estimated number of species: 300

lated mud-loving beetles. The minute mud-loving beetles are small and rarely seen. The variegated mud-loving beetles, or heterocerids (*hetero* = different; *ceri* = horn), are larger, more common, and, as their name implies, have varied coloring: splotches of dull yellow on a background of brown or black. Both the adult beetles and the larvae dig long, winding burrows in the soil. Like moles they often burrow just beneath the surface, heaving the soil as they go and leaving tiny ridges to mark their travels. Sometimes the beetles also erect tiny mud chimneys along their ridges (fig. 136).

Not only are their front pair of legs broad and flat and bordered by spines, but their other two pairs of legs are also beautifully designed for digging. The broad, flat, and powerful mandibles of the mud-loving beetle likewise share in the digging and quicken the pace of excavation. As they dig, the beetles scavenge on the remains of plants and animals that have washed ashore or ended their days on the water's edge. Considering their mucky environments, it is surprising that mud-loving beetles always remain clean and dry. Each beetle is covered with dense silky hair that repels both water and mud. Getting soaked by water, sinking, or getting mired in the mud is physically impossible for these beetles.

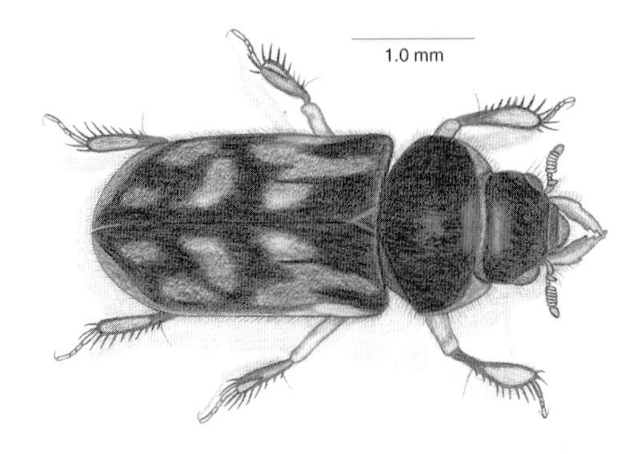

136. The mud around ponds and streams can be thickly populated and thoroughly excavated by mud-loving beetles.

These beetles are clearly beautifully adapted to life in the mud and can be extremely abundant on the banks of streams and the shores of ponds. One way to find out if there are any mud-loving beetles about is to splash water over the mud by the water's edge. When their burrows are flooded, they quickly head for higher ground and often take to the air.

27. Fungus Beetles

Phylum Arthropoda	**Place in food web:** fungivores
Class Insecta	**Impact on gardens:** allies
Order Coleoptera	**Size:** 1–22 mm in length
Around 25 families of beetles have members that live and feed on fungi.	**Estimated number of species:** 79,000

More species of beetles (order Coleoptera) inhabit our planet than do species of any other order of animals; and among the most diverse of these are the beetles that live on fungi. Many fungus beetles are unique enough to merit families of their own. There are minute fungus beetles (Corylophidae), handsome fungus beetles (Endomychidae), hairy fungus beetles (Mycetophagidae), pleasing fungus beetles (Erotylidae), dry fungus beetles (Sphindidae), shining fungus beetles (Scaphidiidae), round fungus beetles (Leiodidae), tooth-necked fungus beetles (Derodontidae), silken fungus beetles (Cryptophagidae), and minute tree-fungus beetles (Ciidae). These families of very different fungus beetles do not even contain all the beetles that dwell on or feed on fungi of the soil (fig. 137).

More than 150 families of beetles are found worldwide. In addition to the families listed in the paragraph above, some of the other families of beetles that have established partnerships with fungi include some mentioned earlier in this book: featherwing beetles, sap beetles, hister beetles, short-winged mold beetles, scarab beetles, and darkling beetles. Not only do beetles feed on fungi, but they also harbor smaller fungi within their digestive tracts. Apparently beetles and the fungi of their guts help each other with digestion and nutrition. Many beetles transport spores of fungi, while butterflies and bees transport the pollen of flowering plants. These ancient partnerships have molded the lives and forms of insects, flowers, and fungi.

137. A handsome fungus beetle is only one of many beetles that spend their larval and adult lives in the company of mushrooms and other fungi.

28. Scorpionflies

Phylum Arthropoda	**Place in food web:** decomposers, detritivores, scavengers, predators
Class Insecta	**Impact on gardens:** absent
Order Mecoptera	**Size:** mature larvae 10–15 mm in length
Families Bittacidae (hangingflies) and **Panorpidae** (common scorpionflies)	**Estimated number of species:** 400

Scorpionflies spend their days and nights in the shadows and thickets of woodland ravines. They begin life in the soil as eggs, larvae, and pupae; only as adults do they leave the soil behind. To anyone who has seen how the males of the family Panorpidae go around with the tips of their abdomens poised like the tails of scorpions, the name "scorpionfly" for this group of insects is obvious. The males cannot sting with their tails; but they do use them to maintain a viselike grip on females during mating (fig. 138).

Hangingflies, the other common family (Bittacidae) of scorpionflies, are often mistaken for the more common crane flies. Crane flies are true flies, and hangingflies are not. Hangingflies and crane flies both have long, spindly legs; a hangingfly has four wings and jaws at the end of its snout. A crane fly, like other true flies, has just two wings and no jaws for chewing, only piercing stylets for retrieving nectar. The hangingfly drapes itself across an insect flyway wait-

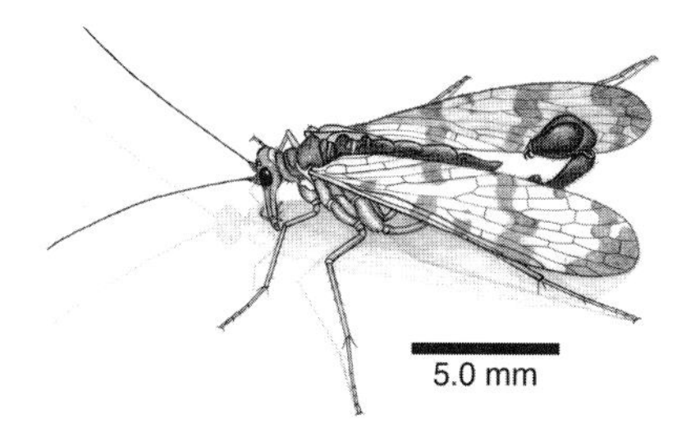

138. Male scorpionflies of the family Panorpidae use their scorpion-like tails for mating, not stinging.

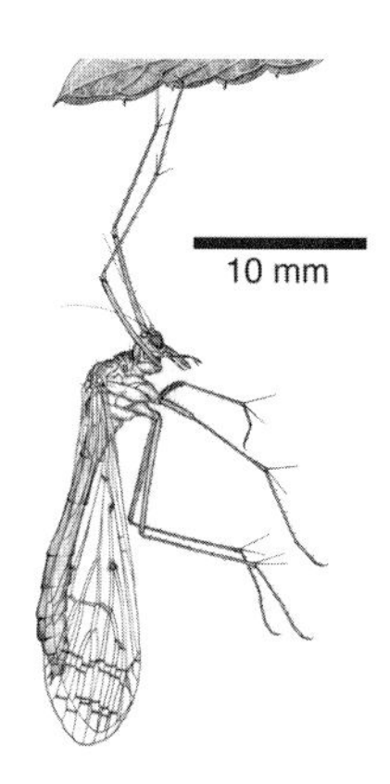

139. This scorpionfly that hangs from vegetation spreads its legs and awaits the passage of a flying insect. Their characteristic pose has earned these gangly insects the name hangingflies.

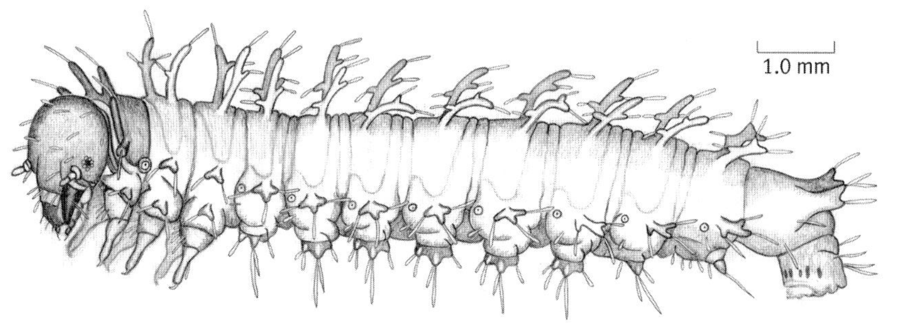

140. The larva of the scorpionfly could be easily mistaken for a caterpillar except that the cuticle covering the front of its head is smooth and does not have the lines that divide a caterpillar's head into distinct regions.

ing for whatever insects fly by. The two front legs of the hangingfly firmly grip an overhanging leaf or twig while its other four legs remain poised to snap shut on whatever insect victim happens by. The long legs close like a clamp and then bring the struggling prey to its jaws (fig. 139).

Covered with bumps, spines, and pieces of dirt, scorpionfly larvae scour the litter and surface of the soil for dead worms and insects. In the leaf litter of the forest, one animal's corpse quickly becomes another animal's meal. Here recycling is not optional but a matter of survival (fig. 140).

29. Antlions

Phylum Arthropoda	**Place in food web:** predators, diggers
Class Insecta	**Impact on gardens:** absent
Order Neuroptera	**Size:** mature larvae 10–22 mm in length
Family Myrmeleontidae	**Estimated number of species:** 2,000

As an antlion flutters by on a hot summer evening, there is little to hint that this delicate, airborne creature was only a few days earlier a ferocious-looking earthbound larva. Its family name, Myrmeleontidae, reflects the fierce nature of its larvae (*myrmex* = ant; *leo* = lion). The best known of these antlion larvae are the pit diggers. Since digging a pit takes a good investment of the antlion larva's energy, the mother antlion has enough forethought to place her eggs in soft soil or sand. The soft, dry soil beneath overhanging cliffs or under the eaves of roofs are often favored spots for egg laying. The newly hatched larva immediately begins construction of the pit that it continually expands and remodels as it molts. For the rest of its larval life, with only its jaws protruding, each larva lies buried in the soil at the bottom of its pit (figs. 141–42).

Insects that chance upon the edge of the pit all too often slip inexorably down the slope into the jaws of the antlion larva. If for any reason an insect's descent into the pit is interrupted, the antlion will facilitate its fall by flicking sand or soil at the prey. Escape is very unlikely. Once the prey is in the grip of its sickle-shaped jaws, the antlion secretes digestive juices that pass through its hollow jaws and begin digesting the prey. The two parts of each jaw, the smaller maxilla and the larger mandible snugly interlock to form the hollow portion of the jaw, with the maxilla overlying a deep groove in the mandible. A few minutes after the digestive juices have been at work on the prey, the antlion larva can begin sucking nutrients from its victim.

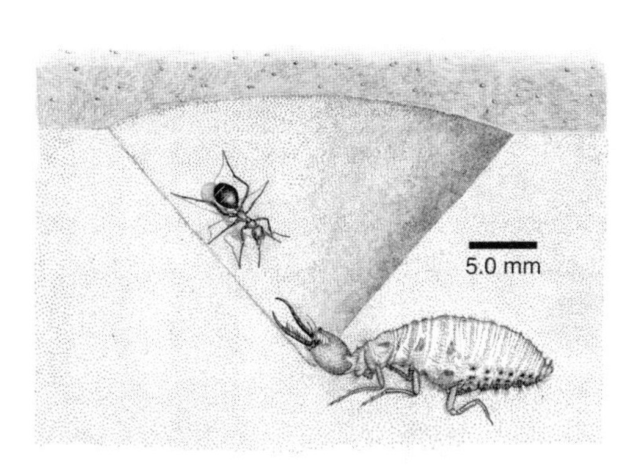

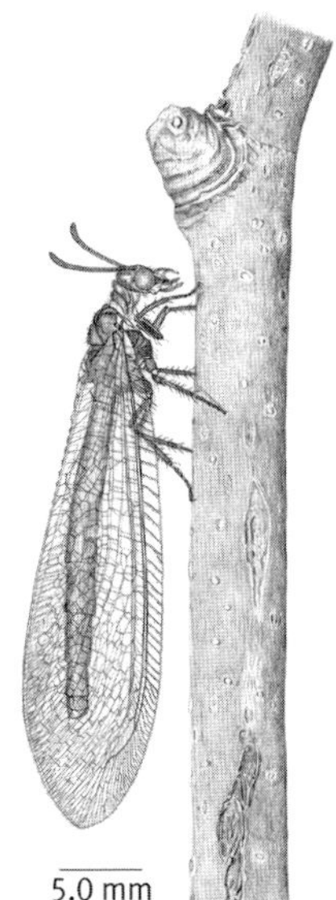

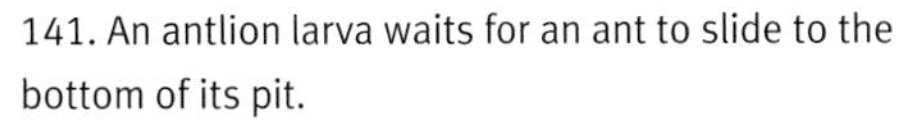

141. An antlion larva waits for an ant to slide to the bottom of its pit.

142. An adult antlion perches on a sycamore branch.

The body of an antlion larva is covered from head to tarsal claws with thousands of sensory bristles that are sensitive to even the slightest vibrations of the soil. An antlion larva is aware that some creature is approaching well before the prey reaches the rim of its pit. By capitalizing on their uncanny abilities to detect nearby movements, larvae of most species of antlions do not even bother to excavate pits. Instead they can be found patrolling runways on the surface or just beneath the surface of a sandy soil, waiting for some insects or spiders to cross their paths.

30. Caterpillars and Moths

Phylum Arthropoda
Class Insecta
Order Lepidoptera
Place in food web: decomposers, detritivores, herbivores, scavengers, fungivores, predators
Impact on gardens: absent/adversaries
Size: mature larvae 6–50 mm in length

Family Tineidae (clothes moth and relatives)
Estimated number of species: 3,000

Family Noctuidae (cutworms)
Estimated number of species: 20,000

Family Psychidae (bagworms)
Estimated number of species: 600

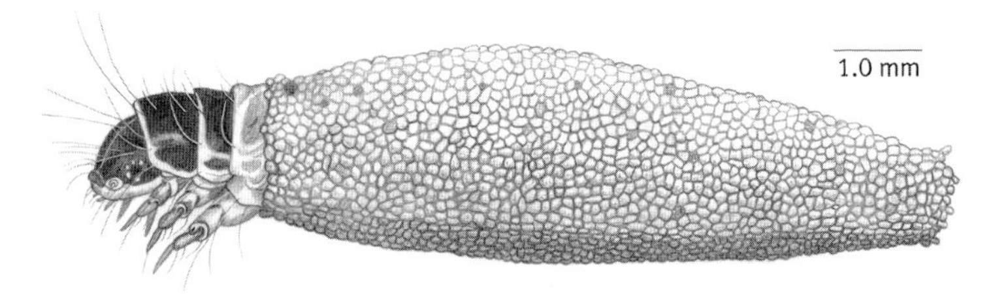

143. This bagworm caterpillar fashions its portable home from particles of the soil as well as with silk from glands near its jaws.

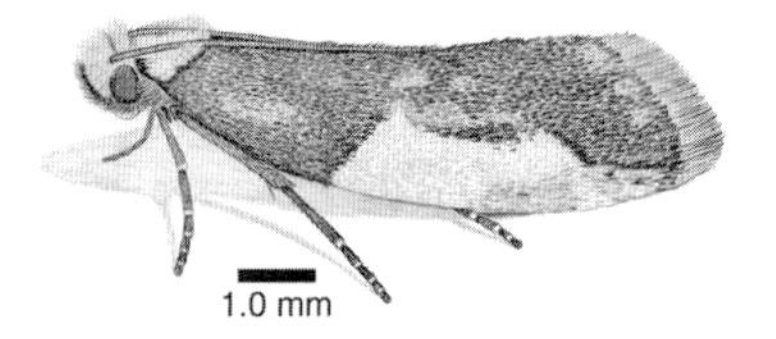

144. A tineid moth begins life in the leaf litter as a caterpillar living in a silken bag and feeding on fungi.

Most of us do not think of those insects with scaly wings, the butterflies and moths, as living in the soil; but a few of them actually do live there as caterpillars. Very few of these caterpillars ever turn into butterflies; most turn into a number of different moths. Although most caterpillars eat leaves of green plants, the few that are soil dwellers survive on roots, decaying plants, or even ants. Some bore into the underground portions of plants, where they live and eat until it is time for them to change into moths. Some live in silk-lined tunnels into which they retreat when they are not feeding on roots. Wherever logs and leaf litter are decomposing, some caterpillars can be found eating the fungi and plant debris. The few butterfly caterpillars found underground live with ants and sometimes feed on the ants (figs. 143–44).

All together there are about 70 families of moths and eight families of butterflies. Most of the soil-dwelling moths belong to one of three families. The same family to which the infamous clothes moth belongs contains many ground-dwelling caterpillars that spin cases of silk and debris, carrying these cases wherever they go. The equally infamous bagworm moth family, which are known to strip entire shrubs of leaves, has many members that begin life in the soil. Each species of bagworm has its own distinctive case made from sand or pieces of plants held together with silk. A few other ground-dwelling caterpillars belong to the family that includes the armyworms and cutworms—caterpillars that are considered serious pests of gardens and crops.

Certainly some of the more eccentric caterpillars live in the soil. Many have forsaken eating fresh, green leaves for brown, decaying leaves; and a few of these ground-dwelling caterpillars have even developed a taste for dung. One desert caterpillar is known to have a preference for the dung of gopher tortoises, and a tropical species feeds exclusively on sloth dung.

31. March Flies, Crane Flies, and Soldier Flies

Phylum Arthropoda
Class Insecta
Order Diptera
Place in food web: decomposers, detritivores, fungivores, algal eaters, herbivores
Impact on gardens: allies/adversaries

March Flies
Family Bibionidae
Size: mature larvae 6–25 mm in length
Estimated number of species: 700

Crane Flies
Family Tipulidae
Size: mature larvae 3–60 mm in length
Estimated number of species: 14,000

Soldier Flies
Family Stratiomyidae
Size: mature larvae 5–35 mm in length
Estimated number of species: 2,000

In most soils and leaf litter, fly larvae can often outnumber any other group of insects, even the very abundant beetles. Like beetle larvae, the larvae of flies have adapted to a wide range of diets—decaying leaves, fungi, algae, and other insects.

March flies are not only most abundant at flowers in March and other months of spring, but they also have a tendency to place all of their eggs in a single batch rather than scattering them over the ground. Each female can lay 200–300 eggs. March fly larvae may number 3,000–12,000 per square meter in one patch of ground, yet be absent only a few meters away. A few March fly larvae may chew on roots of living plants and are considered pests; but most species chew on dead parts of plants and are considered important recyclers (fig. 145).

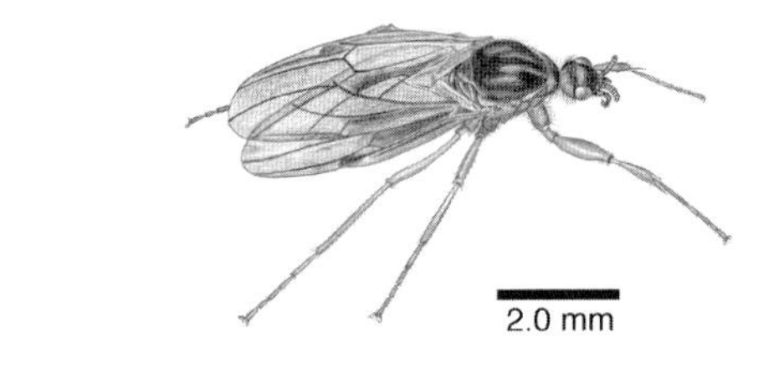

145. In the American Midwest this March fly is more common in April than it is in March. Vast numbers of these flies can emerge from the soil where they have spent the previous year as larvae.

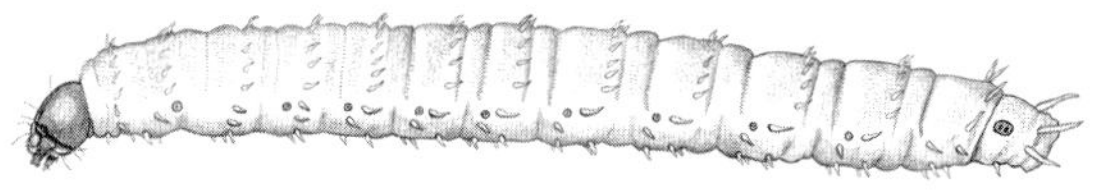

Most people have the impression that all fly larvae look more or less the same but a quick comparison of the larvae of the March fly, the crane fly, or tipulid, and the soldier fly, or stratiomyid (*stratio* = soldier; *myia* = fly), points out some obvious differences. Tipulid larvae usually have fleshy lobes only on their rear segment, while larvae of March flies often have fleshy lobes on all body segments behind their heads. The flat larvae of soldier flies have none of these lobes, but they do have a very tough skin covered with calcium deposits. Tipulid larvae and stratiomyid larvae have two conspicuous spiracles for breathing on their rear segment, but larvae of March flies have spiracles on all but one segment of the thorax and all but one segment of the abdomen. After seeing them a few times, these larvae are easy to distinguish from each other.

Judging from the large numbers of tipulid larvae as well as the numerous stratiomyid larvae that are usually found in samples of leaf litter, one can infer that these fly larvae must be influential recyclers of decaying leaves. Crane flies make up the largest family of flies with 14,000 species worldwide and about a tenth of these are found in the United States. Soldier flies with 2,000 species found around the world and about 300 found in North America are still considered a large family (figs. 146–47).

A tipulid larva has rasping jaws and a head that can be retracted into its thorax. Most larvae use their jaws to chew on moldy leaves, but a few tipulid larvae known as leatherjackets have acquired a taste for roots of crop plants. At least in Europe, leatherjackets have been a considerable annoyance to farmers.

The tough skins of soldier fly larvae are covered with a few long

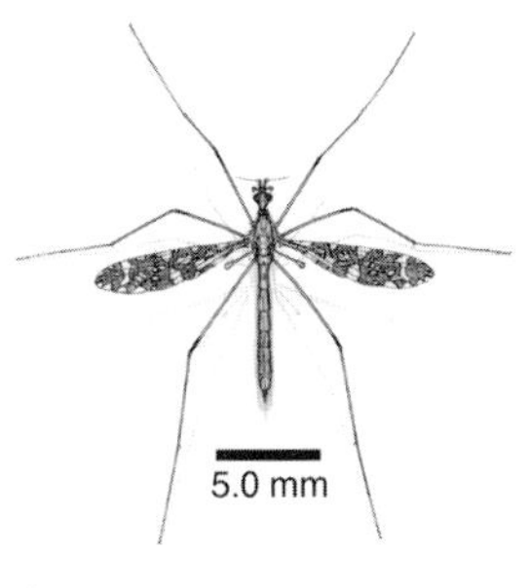

146–47. Crane flies (146) gracefully maneuver their long, lanky legs as they fly through the forest undergrowth. Their larvae (147), however, are completely legless and have only short appendages at their tail ends.

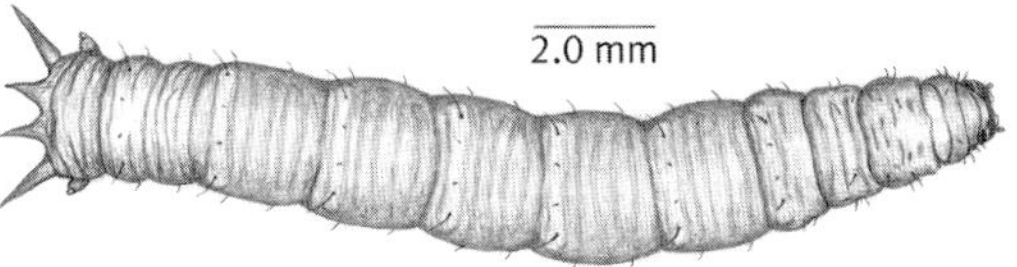

bristles and are reinforced with thousands of grains of calcium carbonate. These circular grains measure about 1/200 of a millimeter across and are arranged in highly ordered geometric arrays over the entire surface of a larva. With such impervious armor, the larvae can survive the summer droughts that often strike even moist pockets of leaf litter. Although the soil algae and fungi that the larvae often swallow along with decaying leaves may wither and perish, stratiomyid larvae endure until the rains come again.

After metamorphosis, life changes drastically for fly larvae. Tipulids transform into thin flies with long, spindly legs that are often mistaken for overgrown mosquitoes or certain scorpionflies. The family name of Tipulidae is derived from a Latin word for spider (*tipula*) and no doubt refers to those long, ungainly legs. It is hard to imagine that some of these delicate flies in their younger days were tough, voracious larvae known as leatherjackets. These flies have exchanged the hard chewing mandibles of their larval days for delicate stylets that sip nectar and feed at sap flows.

Adult soldier flies give up life in the leaf litter to spread their wings and fly among flowers. They are nectar feeders, and some are uncanny impersonators of wasps. Those of one species, with their metallic blue color and white-stockinged feet, look just like mud dauber wasps (figs. 148–49). After being stung once, most birds know enough to stay clear of wasps and their stingers; any insect that looks like a wasp is given the same respect.

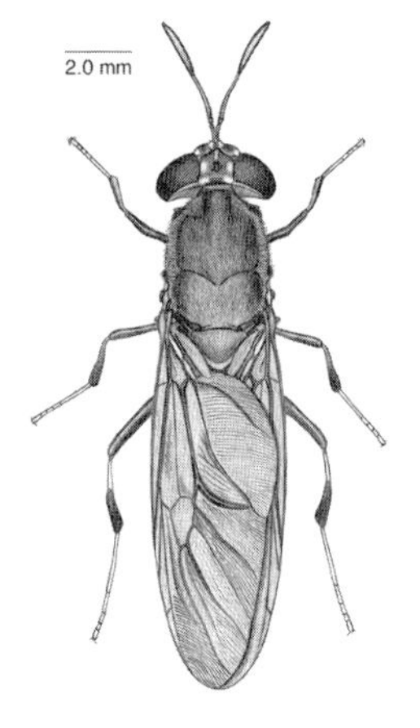

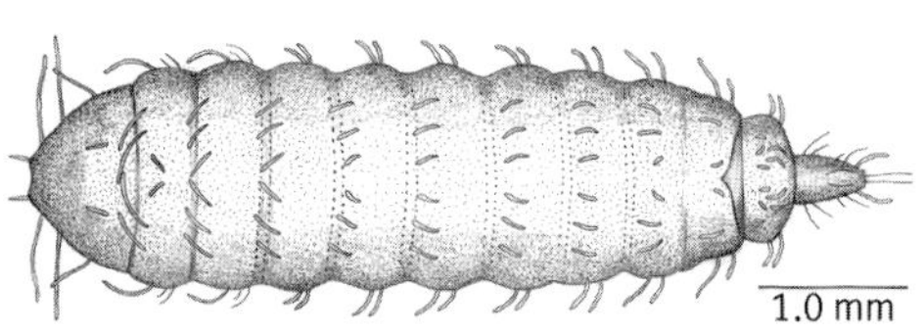

148–49. This soldier fly (148) is a common visitor to compost piles where its larva (149) is a common resident. It is such a good mimic of a mud dauber wasp that birds avoid it rather than risk the possibility of being stung.

32. Midges and Biting Midges

Phylum Arthropoda
Class Insecta
Order Diptera
Place in food web: decomposers, detritivores, fungivores, algal eaters
Impact on gardens: allies

Midges
Family Chironomidae
Size: mature larvae 1.5–30 mm in length
Estimated number of species: 5,000

Biting Midges
Family Ceratopogonidae
Size: mature larvae 2–10 mm in length
Estimated number of species: 1,200

Larvae of midges are considered useful indicators of water quality in flowing waters. Some species can tolerate polluted environments while others are found only in unpolluted habitats. The presence of certain midge larvae in a soil is probably also a good indication of soil quality. Species of soil-dwelling midges are not as numerous as those of ponds and streams, nor are they as well studied. Like their aquatic relatives, however, soil midges recycle plant debris and feed on the tiny algae that are scattered there.

Midge larvae, like all fly larvae, do not have true jointed legs, but they do have two pairs of fleshy prolegs without any joints whatsoever. One of these pairs projects beneath and behind the head of the larvae, and the other pair is found on the very last segment. The tip of each proleg is crowned by a circle of curved spines. These spines and

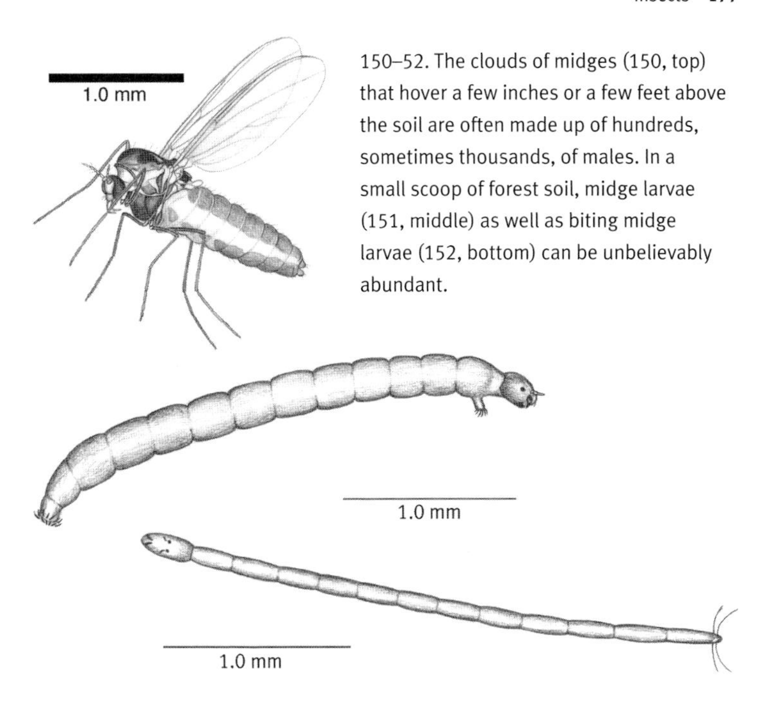

150–52. The clouds of midges (150, top) that hover a few inches or a few feet above the soil are often made up of hundreds, sometimes thousands, of males. In a small scoop of forest soil, midge larvae (151, middle) as well as biting midge larvae (152, bottom) can be unbelievably abundant.

prolegs probably provide enough traction for the larvae to wriggle their way through the leaf litter.

Small clouds of adult midges often hover just above the ground where they once lived as larvae (figs. 150–52). An insect net swept through one of these swarms almost always catches only male midges. These are courtship swarms that males initiate. A female periodically flies into the swarm, and one of the males immediately claims her as a mating partner. There is little time and little place for courtship in midge society.

Midges of the family Chironomidae have some infamous relatives in the smaller family Ceratopogonidae, known appropriately as biting midges, no-see-ums, or punkies. These minuscule flies not only annoy us humans with their blood-sucking habits, but they also take blood from the wing veins of larger insects like moths, dragonflies, and beetles. Biting midges may not feast as well on insect blood as they do on human blood, but at least they do not have to worry about being swatted; they may be smaller than a millimeter, but their bites are mighty and out of proportion to their size.

33. Moth Flies

Phylum Arthropoda
Class Insecta
Order Diptera
Family Psychodidae
Place in food web: decomposers, detritivores, scavengers, fungivores, algal eaters
Impact on gardens: allies
Size: mature larvae 3–6 mm in length
Estimated number of species: 450

Most of us are acquainted with the fuzzy moth flies that parade at times on the edges of our toilet sinks and the walls of our showers. They actually look like tiny moths with exceptionally long antennae and fidgety movements. The country relatives of our household moth flies live in leaf litter and are far less conspicuous, but nevertheless they are down there at all seasons, scavenging decaying plant matter along with algae and fungi.

Some of these moth fly larvae, like the one pictured here, don chunks of soil upon the long bristles that cover their backs. Their earthy attire probably serves as an effective camouflage and protection, at least from us humans, for the only time you are likely to spot any of these earth-colored larvae in soil and leaf litter is when they happen to fall into a white-bottomed collecting jar (figs. 153–54).

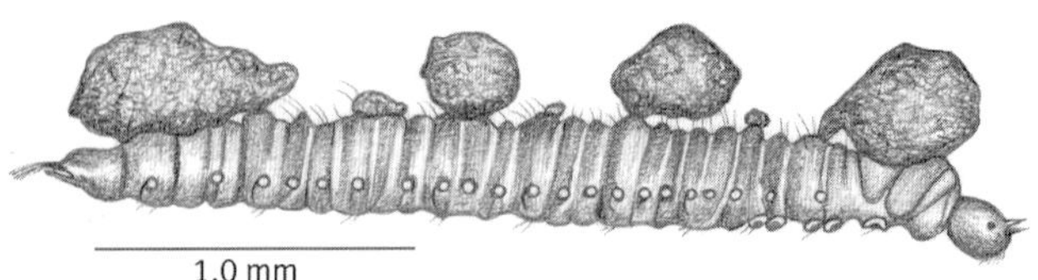

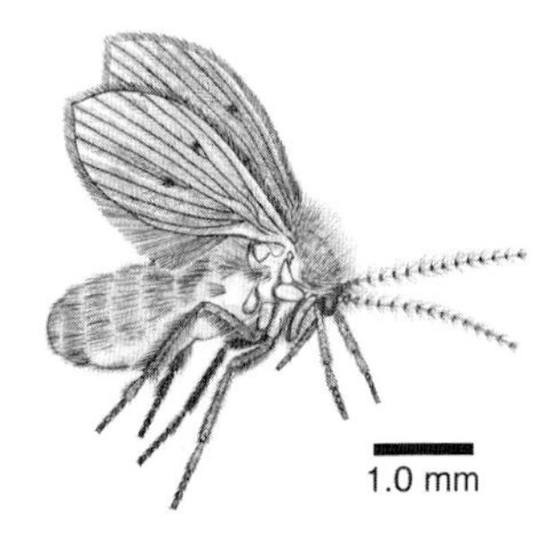

153–54. Larvae of moth flies (153, top) can be abundant in moist, often wet, soil and leaf litter. This moth fly (154, right) grew up in the soil of an oak forest, but a few desert species live in animal burrows. Most adult moth flies survive on sugary or decaying plant material; but in the tropics and around the Mediterranean, some flies are blood suckers and carriers of human diseases.

34. Snipe Flies

Phylum Arthropoda	**Place in food web:** predators
Class Insecta	**Impact on gardens:** absent
Order Diptera	**Size:** mature larvae 6–15 mm in length
Family Rhagionidae	**Estimated number of species:** 400

In late spring and early summer, snipe flies leave the forest litter where they have lived for most of the past year. They search for perches on plants of the forest understory and there await the arrival of passersby that are slow enough for them to catch and eat. A common and lovely snipe fly by the name of *Chrysopilus* (*chryso* = gold; *pilus* = cap) is a fly that stands out among the flies. The iridescent disc of gold scales on its thorax and the patches of white scales on its black abdomen impart a singular beauty to this fly with dappled wings. But its stay aboveground is an ephemeral one. By the end of June few snipe flies remain to cruise the understory of the forest. They have laid their eggs in the soil litter, and their larvae will prey on other insects of the leaf litter until the following spring, when they grow their first legs, spread their wings, and take to the air (plate 32).

Some close relatives of snipe flies live in sandy or dusty habitats, where the larvae lie at the bottom of pits, waiting until some insect or spider chances by and slips into their jaws. This lifestyle may sound very familiar. Because larvae of antlions use a similar strategy to capture prey, the fly larvae are by analogy called wormlions. Their scientific names were also chosen to reflect their parallel lives. Antlions belong to the family Myrmeleontidae (*myrmex* = ant; *leo* = lion), and wormlions belong to the family Vermileonidae (*vermi* = worm; *leo* = lion). Until recently wormlions were considered eccentric members of the snipe fly family, but now they have been assigned to this small family of their own with about 27 species of flies found in dusty places around the world.

35. Robber Flies

Later in the summer, robber flies take up some of the same posts that were occupied by snipe flies. Robber flies continually dash off in pursuit of passing insects but return again and again to these favorite posts.

These flies have all the features an effective predator should have.

Phylum Arthropoda	**Place in food web:** predators
Class Insecta	**Impact on gardens:** allies
Order Diptera	**Size:** mature larvae 8–35 mm in length
Family Asilidae	**Estimated number of species:** 5,000

With their long, sturdy legs and big claws, they do not hesitate to pounce on insects just as large or even larger than they are. Their bite is quick and potent, as anyone who has tackled with one of these flies knows. Their large, bulging eyes endow them with keen vision for discerning shapes and movements.

Coming from a large and diverse family of flies, robber flies also have their differences. Some are slender and streamlined. Many others are stout and fuzzy, sounding and looking like bees. Such behavior may serve them well in sneaking up on their prey as well as protecting them from birds that associate eating a bee with being stung (figs. 155–56).

Larvae of robber flies either live in the soil or in well-decayed logs. Even without the large eyes, sturdy legs, and strong wings of their parents, these larvae obviously manage quite well as predators of other larvae. Their bodies are long, smooth, and tapered at both ends. They can probably move faster than most soil larvae, and they

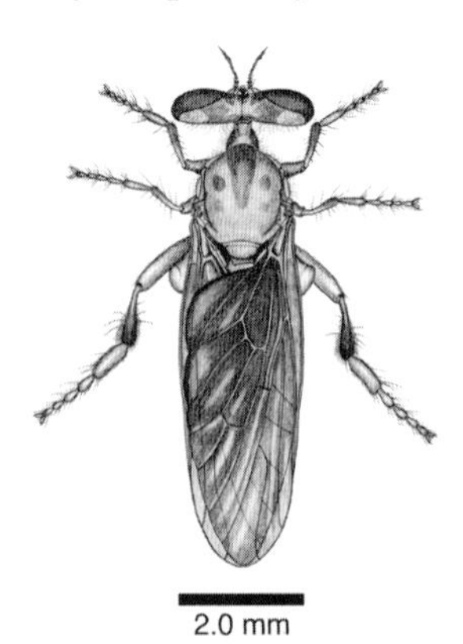

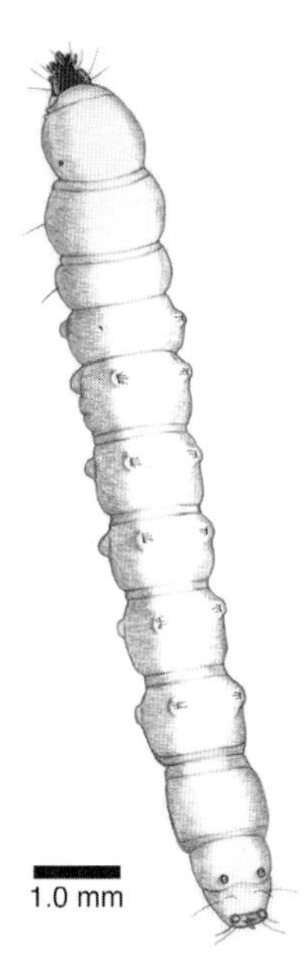

155–56. Robber flies (155, top) can be considered the thugs of the flies, pouncing on prey as large as grasshoppers and dragonflies or as threatening as stinging bees and wasps. The larvae (156, right) will attack beetle grubs that are twice their size.

feed on enough of these larvae to grow as long as an inch (three centimeters) or more.

36. Bee Flies

Phylum Arthropoda	**Place in food web:** parasites
Class Insecta	**Impact on gardens:** allies
Order Diptera	**Size:** mature larvae 9–22 mm in length
Family Bombyliidae	**Estimated number of species:** 3,000

Parasitic larvae of this unusual family of flies have an uncanny ability to find hosts that live underground. The adults are fuzzy like bees, buzz like bees, and are shaped like bees. They even hover and visit flowers like bees. They are among the few insects that visit the very first flowers of spring. The mother bee fly simply drops an egg near the burrow of a likely host, and the lively larva that hatches from the egg goes off on its own in pursuit of its host. Several hundred different species of bee flies live in the United States, and they choose their hosts from among the beetles, flies, moths, antlions, digger bees, and grasshopper eggs that are found in the soil. Some bee fly larvae specialize in finding antlion larvae, others feed only on grasshopper eggs, and one common species is a parasite of the ferocious-looking larvae of tiger beetles. Having expended a great deal of energy finding

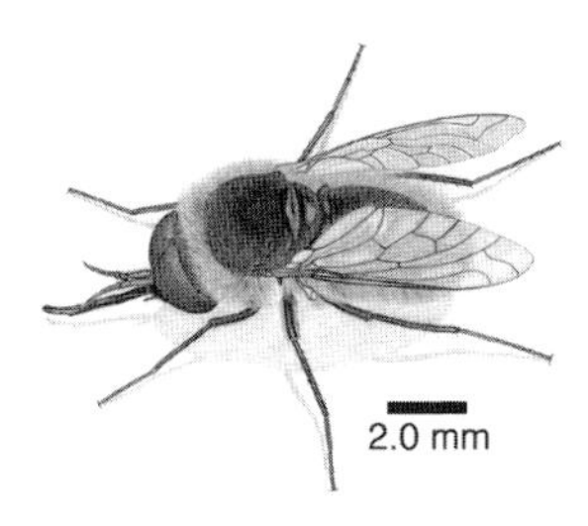

157–58. The bee fly (157, top) may look like a bee and buzz like a bee, but it neither stings nor bites. Its menacing-looking proboscis is used only for sipping nectar. Its larva (158, right) is a sluggish parasite of other insect larvae and pupae that live underground.

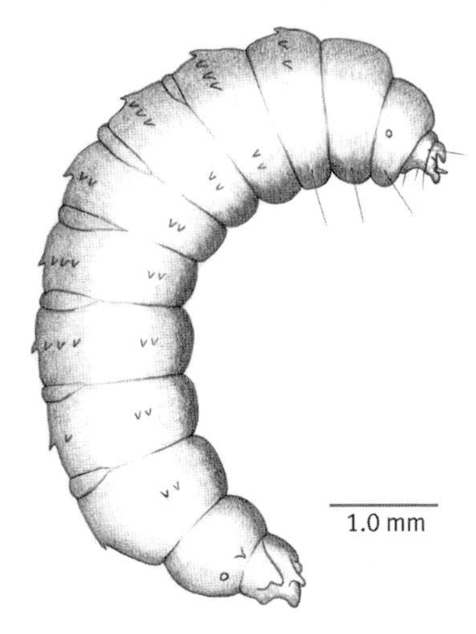

a host, the young bee fly larva settles down and molts into a lethargic larva that grows at its host's expense (figs. 157–58).

37. Long-legged Flies

Phylum Arthropoda
Class Insecta
Order Diptera
Family Dolichopodidae
Place in food web: decomposers, detritivores, predators

Impact on gardens: allies
Size: mature larvae 6–22 mm in length
Estimated number of species: 6,000

These sprightly flies prance about on vegetation with their long and spindly legs. Their family name, Dolichopodidae (*dolicho* = long; *podi* = foot, leg), conveys just one of the many striking features of these handsome little flies. They have long legs as well as long feet, or tarsi, as feet are known in the insect world. Both males and females are iridescent green and blue. As if their lustrous colors were not sufficient embellishment, male flies have a variety of other adornments that impress female flies looking for just the right mates. During courtship, males show off the ornaments on their legs or antennae as well as their large, conspicuous genitalia. Some insects may lead retiring lives and are rarely observed, but dolichopodid flies are commonly seen and are rarely forgotten (plate 33).

As predators of smaller flies, the adults move in quickly to attack and feed. Like the larvae of all flies, larvae of long-legged flies are legless, but they have streamlined forms and wriggle gracefully through the humus, decaying leaves, and rotting leaves that they call home. Far more is known about the colorful adults than the larvae that travel the passageways of the leaf litter. Like the adult flies, some larvae may be hunters of other insects, but probably most dolichopodid larvae graze on decaying plant matter, helping to break it down into the reserves of humus that characterize healthy soils.

38. Picture-winged Flies

Root nodules are remarkable resources that nourish plants and bacteria, as well as certain animals of the soil that have discovered these nodules to be some of the most nutritious portions of plants. Even though plant tissues are always rich in sugars and other carbohy-

Phylum Arthropoda	**Place in food web:** herbivores
Class Insecta	**Impact on gardens:** adversaries
Order Diptera	**Size:** mature larvae 4–9 mm in length
Family Platystomatidae	**Estimated number of species:** 1,000

drates, nodules of plants contain all these nutrients as well as millions of rhizobial bacteria. Certain enterprising maggots of one fly family have developed a special relationship with rhizobia and root nodules. Maggots of platystomatid (*platy* = flat; *stomato* = mouth), or picture-winged, flies may look like ordinary maggots, but they do not behave like ordinary maggots. Environments like decaying flesh or rotting leaves are the usual haunts of most maggots; but rather than scavenge in such places, platystomatid maggots have acquired distinctive and relatively refined tastes. Newly hatched maggots navigate through the soil beneath plants with root nodules and survive by eating these nodules. Since a single plant can have hundreds of nodules on its roots, these maggots always have an abundant source of nutrient-packed food. What is surprising is that more creatures have not discovered this resource of the soil (figs. 159–60).

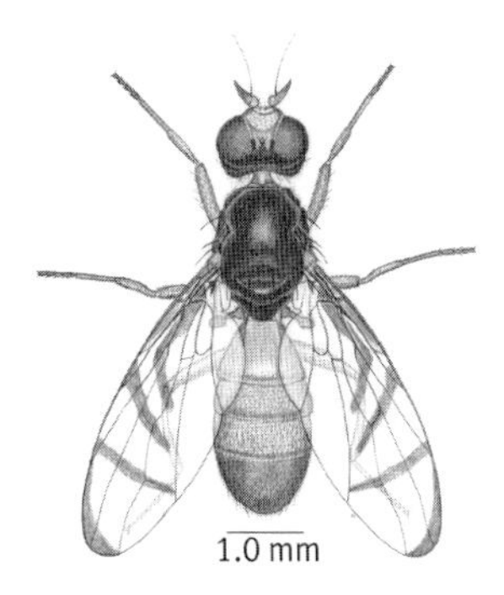

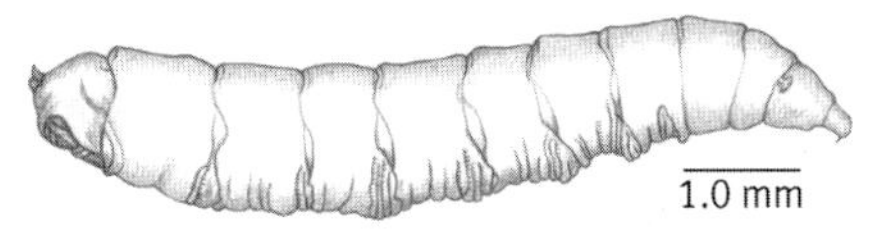

159–60. Picture-winged flies (159, left) have developed a special relationship with root nodules. Rhizobial bacteria of root nodules not only enrich the soil by converting dinitrogen gas from the air to ammonia, but they also nourish the larvae (160, top) of these flies.

39. Root-maggot Flies

The root-maggot flies are the most notorious members of a family of flies whose members are closely related to the common house fly. At close range, what stands out about these flies are their bright red eyes and the long, curved spines that cover their bodies. While root-maggot flies feed on roots of vegetables, many other members of the family are dung feeders. Some of these dung feeders prefer dung that

Phylum Arthropoda
Class Insecta
Order Diptera
Family Anthomyiidae
Place in food web: herbivores, scavengers, coprophages, parasites
Impact on gardens: adversaries/allies
Size: mature larvae, 4–12 mm in length
Estimated number of species: 1,100

they find in burrows of tortoises and rodents. Other members of this family are parasites of grasshoppers or digger bees, and some scavenge in bird nests or on dead animals.

Root maggots spend their winters in garden soils and are most active during the cool days of spring and fall. Not until the longer days of spring do they pupate; and soon thereafter the flies emerge from the soil. Cabbage maggots, turnip maggots, seedcorn maggots, and onion maggots are all species of root maggots that feed on germinating seeds of corn, beans, and peas, as well as on roots of vegetables in the cabbage family and bulbs of members of the onion family (figs. 161–62).

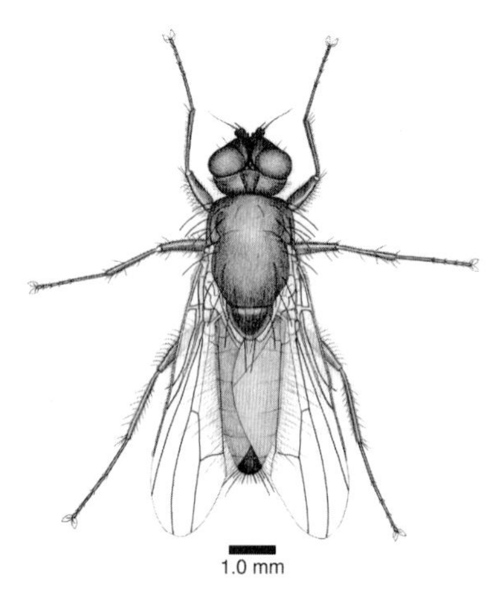

161. Onion odors attract this red-eyed fly to gardens. The onion maggot fly is found throughout the Northern Hemisphere, probably introduced to the northern United States and southern Canada from northern Europe during the nineteenth century.

162. Cabbage maggots burrow into roots and stalks of members of the cabbage family. In fields and gardens with thriving communities of soil creatures, root maggots have many natural enemies that keep their populations in check.

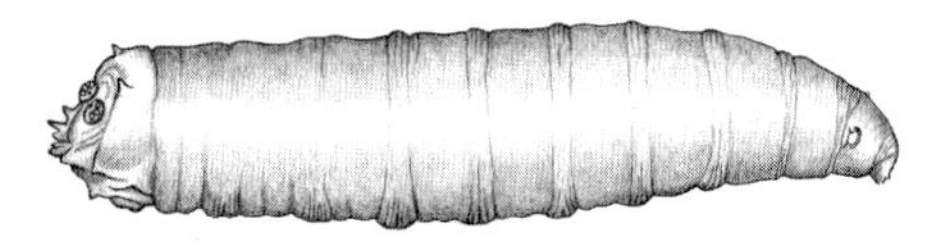

To control populations of these root feeders, farmers and gardeners have recruited natural enemies of root maggots. These are certain genera (*Metarhizium, Beauveria*) of soil fungi in the phylum Ascomycota that, depending on environmental conditions, alternate between feeding on decaying plant matter and surviving as predators of soil-dwelling insects. These fungi are endowed with an entire constellation of enzymes and other chemicals for processing their diverse food, including those that digest insect cuticles and some that act as potent insecticides. To see if these fungi inhabit your garden, place

---

THE DIVERSITY OF DECOMPOSERS AMONG MAGGOTS

Diptera, the order of insects represented by the true flies, contains three suborders and about 108 families. The previous two fly families—Platystomatidae and Anthomyiidae—are just two of around 60 families in the suborder of the Diptera known as Cyclorrhapha (*cyclor* = circular; *rhapha* = suture). Some of the better-known adult members of this suborder of flies include the familiar house fly, the blow fly, and the fruit fly; their larvae are the familiar maggots. The name of their suborder is based on a unique feature of these flies' metamorphosis. At pupation the larval skin of these flies turns brown and hardens to form a protective cover, or puparium (*pupa* = pupa; *-arium* = chamber), for the soft pupa. At the head end of the puparium, where the adult fly will eventually emerge, a weak circular line, or suture, forms in the otherwise uniformly hard puparium. When the fly that forms within is ready to emerge, it inflates a sac on top of its head, between its two antennae, pushing off the circular "cap" outlined by the suture.

Maggots from many families in this suborder are encountered in soil, on fungi, and in decaying vegetation as well as on carcasses and in dung. Hump-backed flies (Phoridae), small dung flies (Sphaeroceridae), black scavenger flies (Sepsidae), and stilt-legged flies (Micropezidae) are members of just a few of the families that can be abundant decomposers and recyclers. These flies are often small and difficult to distinguish from each other; members of some families resemble small house flies both as maggots and as adults. Some of the adult flies found in plant litter are not only tiny but wingless. While most flies leave the soil after their days as maggots are over, these wingless flies remain in the litter alongside their maggots, preferring to feed underground on fungi and decaying vegetation. These flies and their maggots might be small and easily overlooked, but the millions that constantly toil at recycling on a small patch of earth contribute a great deal to soil quality (fig. 163).

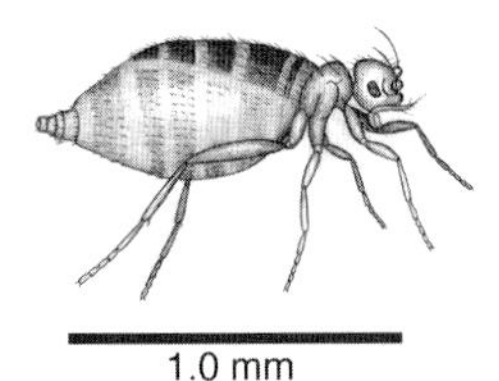

163. This unusual, tiny fly of the unusual family Phoridae lives its entire life in the leaf litter.

some moist soil in a covered dish such as a Petri dish, along with a dead beetle or fly. Then wait to see if the insect's hard cuticle becomes covered with a fuzzy growth of white fungal hyphae.

40. Gall Wasps

Phylum Arthropoda	**Place in food web:** herbivores
Class Insecta	**Impact on gardens:** absent
Order Hymenoptera	**Size:** 1–8 mm in length
Family Cynipidae	**Estimated number of species:** 1,200

On February days, often on snow-covered ground, tiny, stingless wasps crawl out of the leaf litter and begin their long trek from the soil to the treetops. Not many people ever witness the migration of these wingless wasps, which look more like fleas. They have just emerged from chambers, or galls, in the leaf litter where they have passed the autumn and the earlier days of winter. Their lives began the summer before when their mothers laid eggs on leaves or buds of an oak tree. Either the act of the mother wasp's egg laying or some stimulus from the newly hatched larva resulted in the part of the leaf or bud around the egg growing into a gall that sheltered and nourished the young gall wasp; there it remains until the lengthening days of February tell it that the time has come to begin its pilgrimage back to the branches where it had spent its youth. Some galls are plain and simple growths; others are elaborate and ornate (fig. 164).

These little wasps have one of the strangest genealogies you can imagine. All the wasps that emerge from the galls in the leaf litter are virgin females, and they are on their way to lay eggs on the new buds of the oak trees. Here new and different galls will form around the eggs that they lay, and the wasps that issue from these galls look very different from their virgin mothers. In the insect world, virgin

164. In late winter, small, stingless wasps emerge from these galls that formed on oak leaves the summer before and that have lain on the forest floor since autumn. (Michael Jeffords)

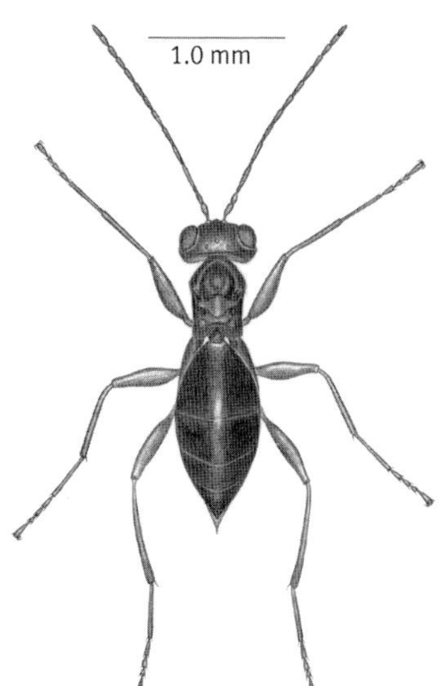

165. This virgin female wasp leaves her gall in the leaf litter and climbs a nearby oak tree to lay her eggs on unopened leaf buds.

births are commonplace events. Even though a wingless female has no males to court her and does not expend any precious energy on mating during the chilly days of February, her children include both males and females (fig. 165).

Unlike their mother's generation, however, all her children have wings, mate, and give birth to another generation of all female wasps. The generation of male and female wasps hatches from unfertilized eggs in late winter or early spring, while the generation of all female wasps hatches from fertilized eggs laid in the summer. The generations of gall wasps alternate, with grandmothers and granddaughters having more in common than mothers and daughters. Fathers never have sons, only grandsons. Mothers always have daughters and granddaughters, but only in alternate generations do they have sons or grandsons. No wonder biologists were puzzled by the strange ways of gall wasps until the German scientist Hermann Adler had the patience in 1875 to rear many of these wasps and to observe carefully the relationships among wasps, their galls, and their sex lives.

41. Parasitic Wasps

Phylum Arthropoda
Class Insecta
Order Hymenoptera
Many families of wasps (~36) have members that are parasitic on other insects.

Place in food web: parasites
Impact on gardens: allies
Size: 1–40 mm in length
Estimated number of species: 100,000

Parasitic wasps scurry about on spindly legs, waving and tapping their antennae as they track down the odors of their hosts. Unlike parasitic tapeworms and roundworms, parasitic wasps choose only other insects, spiders, mites, and a few other arthropods as hosts. All types of arthropods of the leaf litter—predators, decomposers, herbivores, and even other parasitic wasps—can be hosts for the larvae of parasitic wasps. About 15 percent of all insect species are parasitic, and half of the estimated 200,000 species of Hymenoptera (ants, bees, wasps) are parasites. Some wasps are very specific about which hosts they parasitize, and at which stage in the host's life; others are less particular and are content with trusting their eggs to any number of eligible hosts. Many parasitic wasps of the soil, like some tiny gall wasps and tiny flies of the leaf litter, have completely lost their wings. Down in the cramped spaces of the leaf litter and the tortuous passageways of the soil, where mother wasps go in search of the best hosts for their eggs, wings are probably more of a hindrance than an asset (figs. 166–67).

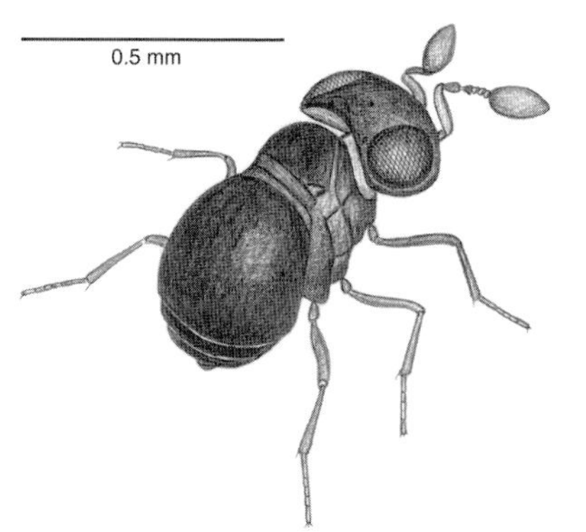

166. This tiny parasitic wasp parasitizes the eggs of other soil arthropods.

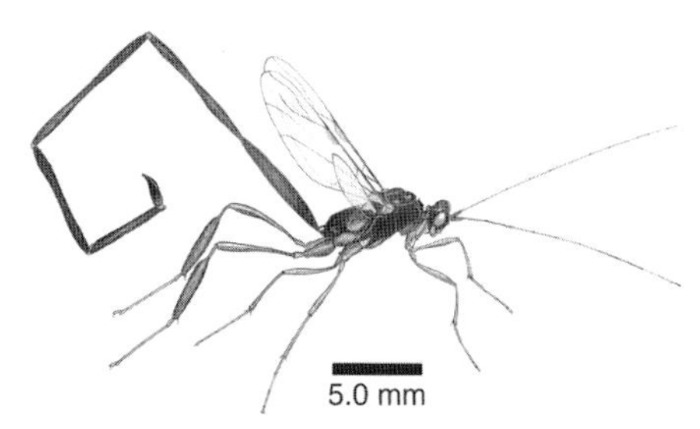

167. This large pelecinid wasp has an exceptionally long abdomen. She uses the ovipositor at the end of the abdomen to probe beneath the soil surface for beetle grubs on which to lay eggs of her parasitic larvae.

42. Digger Bees and Velvet Ants

Phylum Arthropoda
Class Insecta
Order Hymenoptera

Digger Bees
Families Andrenidae, Halictidae, Colletidae, Apidae
Place in food web: diggers, herbivores
Impact on gardens: absent
Size: 5–15 mm in length
Estimated number of species: 20,000

Velvet Ants
Family Mutillidae
Place in food web: parasites of insects
Impact on gardens: absent
Size: 6–20 mm in length
Estimated number of species: 5,000

Before honey bees arrived from the Old World in the 1600s, native bees such as sweat bees and mining bees were the only bees that visited and pollinated the flowers of the Americas. Most of these bees are digger bees that nest in the soil. Even though each bee digs its own burrow, it often digs it only an inch or two (a few centimeters) from the burrow of another bee. So many burrows may exist side by side that a community of digger bees may cover many acres, and as many as one and a half million bees may nest on an acre of land. By far the largest community of digger bees ever recorded was discovered in 1963 along the River Barysh in Russia. There, along a four-mile (six-kilometer) stretch of one river bank 12.2 million nests covered an area of about 90 acres.

These nests of digger bees can take a number of forms, with each species of bee having its own architectural design. Some bees dig burrows only two or three inches (five to eight centimeters) deep whereas other bees may dig as deep as two or three feet (60 to 90 centimeters). Anywhere from one to many brood cells can be arranged either horizontally or vertically around the main shaft of a burrow. Sometimes the brood cells are clustered like grapes at the end of the main shaft; sometimes they are spaced at wider intervals along its length. Within each of the brood cells the mother bee places a mixture of pollen and honey. Just the right amount is added to carry each bee larva through its metamorphosis.

Mining bees of the family Andrenidae leave their burrows in the spring, exactly when the particular pollen and nectar with which they provision their brood cells is available. The life cycle of each bee is intimately coupled to the life cycle of its favorite spring flowers. After mining bees mate and lay their eggs, their larvae grow quickly in the brood cells, transforming to pupae by early summer. Here they wait through the winter for another spring and the arrival of their short-lived host flowers. For some species of bees, no other flowers seem to suffice.

In the spring, sweat bees of the family Halictidae begin appearing as well. All the sweat bees that spent the winter in the soil are females that mated the previous autumn. They begin their life aboveground by digging new brood nests and then stocking them with honey and pollen for their larvae. By midsummer, bees begin emerging from the brood nests, with males appearing first. Vast numbers of males cruise over the thousands of burrows in the community of digger bees, awaiting the first emergence of the females (fig. 168). On a hot July day, one can walk through a town of digger bees and see clouds of males hovering a few inches above the entrances to the burrows. The males are too intent on finding an eligible mate to notice a human intruder. There is an urgency in each male's search for a mate; unlike most females, which will safely spend the coming winter underground, the males will die in a few days.

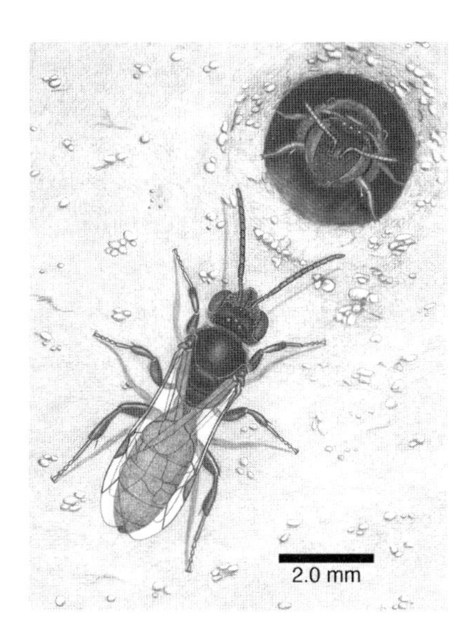

168. A male sweat bee (left) awaits the first appearance of a female from her brood nest.

As the male sweat bees flit overhead, females of a related species of bee may be dashing from burrow to burrow in search of someone else's brood cells in which to lay their eggs. These females are parasitic bees that have come to rely on other bees to dig burrows and gather pollen for their larvae. The parasitic larva has strong jaws with which it can quickly dispatch the rightful resident of the brood cell. It eliminates its nest mate as soon as possible, since just enough pollen is placed in each cell to feed a single larva.

In the cities of digger bees, other insects have found a number of ways to make their livelihoods. Among those insects is a wasp called a velvet ant. Although the male velvet ant has wings but no stinger and looks like a wasp, the female has a nasty stinger but no wings and looks like an ant. Both the male and female velvet ants are covered with combinations of red, orange, yellow, black, and white hairs that give them a velvety appearance. The females spend most of their time moving from burrow to burrow, trying to find unprotected ones in which they can leave their eggs and raise their larvae. Unlike bees that survive as larvae on the pollen of a brood cell, velvet ants are wasps that survive as larvae on the flesh of other insects. The mother velvet ant naturally looks for brood cells with well-developed, pollen-fed bee larvae as homes for her own larvae. She paralyzes a bee larva

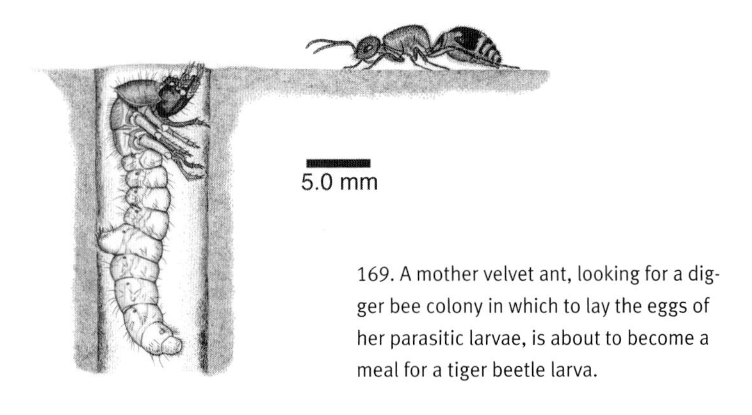

169. A mother velvet ant, looking for a digger bee colony in which to lay the eggs of her parasitic larvae, is about to become a meal for a tiger beetle larva.

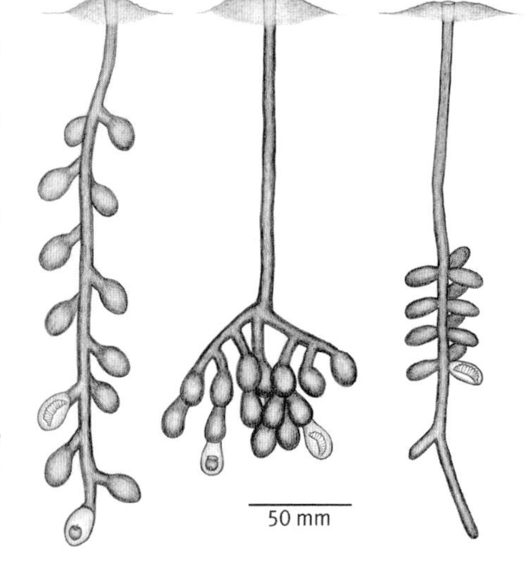

170. The brood nests of different digger bees have their own distinctive architecture. On the left and right are the nests of two different sweat bees, and in the center is the nest of a bee that is a member of the family to which honey bees and bumble bees belong. Some chambers were drawn in cross-section to show either bee larvae or the ball of pollen that serves as a provision for the larvae.

with her stinger, lays an egg and moves on in search of another well-fed larva (fig. 169).

Generation after generation, year after year, a city of digger bees manages to survive threats from wasps and other bees. Just imagine how much soil a million of these digger bees can move in a single year on a single acre of land. If each female bee in the city digs a burrow two feet (60 centimeters) deep, or even one foot (30 centimeters) deep, mineral matter and organic matter of the soil are bound to be well circulated (fig. 170).

43. Digger Wasps

Phylum Arthropoda
Class Insecta
Order Hymenoptera

Sand Wasps and Cicada Killers
Family Sphecidae
Place in food web: diggers, predators
Impact on gardens: allies
Size: 10–30 mm in length
Estimated number of species: 7,700

Spider Wasps
Family Pompilidae
Place in food web: diggers, predators
Impact on gardens: absent
Size: 15–40 mm in length
Estimated number of species: 2,500

Parasitic Digger Wasps
Place in food web: diggers, parasites
Impact on gardens: allies
Size: 10–30 mm in length
Family Scoliidae
Estimated number of species: 300
Family Tiphiidae
Estimated number of species: 1,500

While digger bees provision the underground chambers for their larvae with pollen and nectar, digger wasps stock their burrows with all types of insects and spiders—grasshoppers, crickets, caterpillars, moths, true bugs, flies, bees, beetles, leafhoppers, treehoppers, even butterflies and trapdoor spiders.

The name digger wasp refers to a number of different wasps, each with its own hunting preference and its own distinctive features. The digger wasps known as thread-waisted wasps have skinny waists where the thorax joins the abdomen, and each of these wasps grows up on a diet of caterpillar, grasshopper, or cricket meat. Spider wasps, as their name implies, prey on spiders, including those that hide behind trap doors. They are long-legged wasps, most of which have dark bodies as well as dark wings. Sand wasps usually have pale green markings and a fondness for just about any reasonably large flies. A single larva of a sand wasp can eat as many as thirty flies in its lifetime, and a mother sand wasp continues bringing flies to its growing larvae even after their growth is well along. Parasitic digger wasps dig through the soil in search of beetle grubs as hosts for their larvae (figs. 171–72, plates 34–35).

Of all the digger wasps, the cicada killer is one of the most striking in terms of size and color. This large black, tan, and yellow wasp is strong enough to subdue a cicada that weighs from four to six times

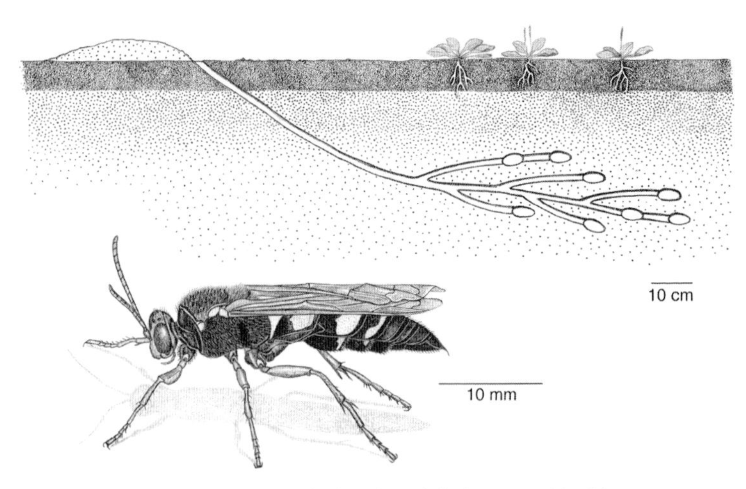

171. Cicada killers provision each chamber of their nests with either one or two paralyzed cicadas.

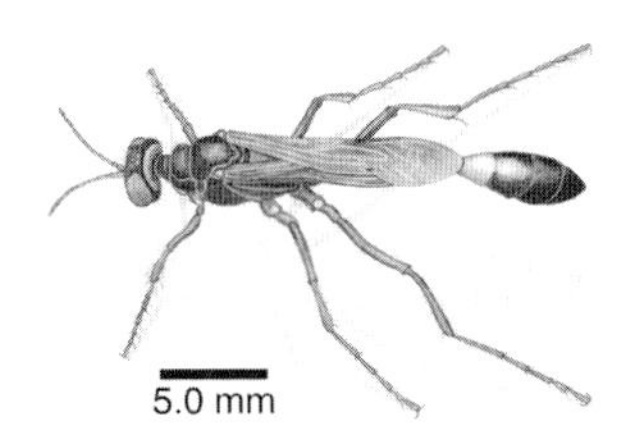

172. Thread-waisted wasps of the genus *Ammophila* (*ammo* = sand; *phila* = love of) carry paralyzed caterpillars into their underground nests.

as much as it does. Once it has paralyzed the cicada with its long sting, the cicada killer somehow manages to maneuver its hefty victim into its burrow or drags the paralyzed cicada up a nearby tree, gaining just enough altitude to glide and fly to its destination.

The underground nurseries that digger wasps prepare for their larvae are simpler and less numerous than the burrows and chambers of digger bees. Each wasp burrow slopes downward to a depth that rarely exceeds a foot (30 centimeters) but that extends in length anywhere from one foot to three feet (30 to 90 centimeters). As predators or parasites on other insects and spiders, digger wasps have limited numbers of edible resources to feed their larvae. Pollen and nectar that bees use for larval food are often present in abundant supply, whereas insects that end up as provisions in the burrows of digger wasps are usually relatively scarce. The land simply cannot support as many carnivores as it can vegetarians. This is true whether the carnivores are lions, tigers, wolves, or digger wasps. Rarely do digger wasp colonies exceed a few hundred burrows, and entrances to these burrows are usually more widely separated than the closely packed entrances in a densely populated colony of digger bees.

Like the males of digger bees, the males in a digger wasp colony typically emerge from their burrows a few days before the females. While awaiting the emergence of the females, the males fly back and forth over the colony. A female rarely has an opportunity to fly from her burrow before a male spots her and immediately mates with her. Males may continue hanging around the colony for weeks. They continually visit flowers for nectar during this time, but they never join in the digging, hunting, and provisioning of burrows that the industrious females so dutifully carry out.

44. Ants

Phylum Arthropoda
Class Insecta
Order Hymenoptera
Family Formicidae
Place in food web: diggers, predators, herbivores, scavengers, fungivores

Impact on gardens: allies
Size: 2–15 mm
Estimated number of species: 8,800

Mites and springtails may be the most common arthropods of soil; but among those arthropods represented by the insects, ants are the most abundant in most soils. All these ants can move substantial amounts of soil and can add substantial amounts of debris to the compost heap of plant and animal remains that surrounds their colony. An ant's reputation for carrying large and heavy objects between its sturdy jaws is legendary and well deserved.

However a person calculates the amount of earth moving that ants are responsible for, the numbers always come out large. Subsoil and the elements that have been carried down to it by rain and melting snow are constantly being moved to the soil's surface during the construction of anthills. The galleries of harvester ants in the deserts of the American Southwest can extend as deep as 15 feet (4.5 meters) underground. The workers of another species of desert ant move 150–300 pounds (70–140 kilograms) of subsoil to the surface each time they build an anthill. Ants in a Massachusetts field moved enough soil

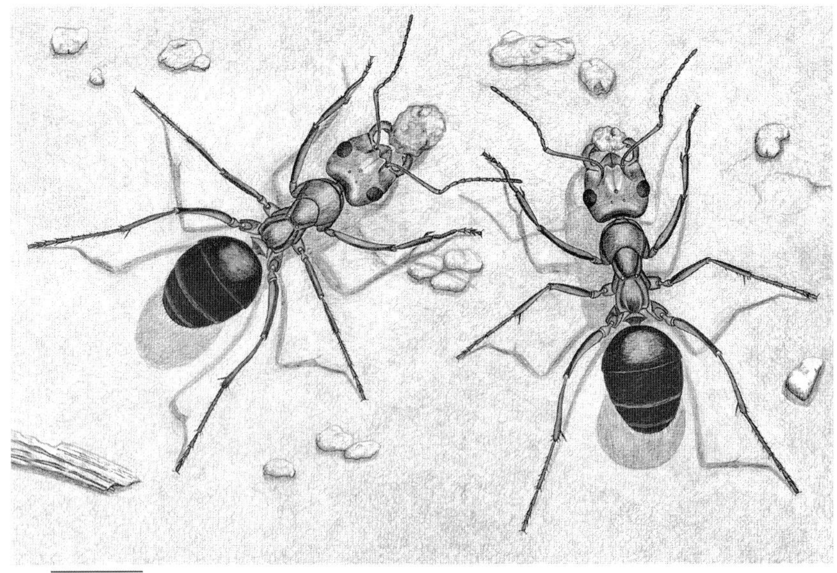

173. Allegheny mound-building ants are notorious warriors as well as movers of earth. Other ant species that enter the neighborhood of their large mounds are soon attacked and decapitated.

in one year to cover the surface with a 0.2 inch (5 millimeter) layer, weighing 30 tons per acre.

The mound-building ants of eastern North America seem to build the most imposing anthills. These insect earthworks are several feet in diameter and often two to three feet (60 to 90 centimeters) high. Generations of ants can occupy a mound, through many summers and many winters, continually expanding their excavations and their composting. Ants range widely from their mounds along well-worn trails, carrying seeds, other insects, and pieces of plants back to their colony, where much of this plant and animal matter is composted along with dead ants and ant droppings (fig. 173, plate 36).

Harvester ants of the desert carry 15 million seeds per year on an acre of land to their underground granaries. Since the total crop of seeds on an acre of desert is about 1.5 billion, the harvester ants add about 1 percent of the seed crop to the soil in the immediate vicinity of their nest.

In the forests of Europe, wood ants keep organic and mineral matter in constant circulation around their nests. Each day during the warmer months of the year the foragers for a nest of about 1 million wood ants bring at least 100,000 insects into their nests as food for the colony.

Higher levels of nutrients derived from decomposition of organic debris naturally accumulate in the soil found around an ant colony than in the soil beyond the ant nest. With the great variety of organic matter added to it, the compost of an ant's nest is especially rich in plant nutrients and provides a rich environment for decomposers as well as a rich nursery for young, delicate seedlings that are setting down their first roots. Ants assure survival not only of the seedlings but also of the animals that have come to depend on these seedlings.

Other relationships have developed between ants and a variety of arthropods appropriately called myrmecophiles (*myrmex* = ant; *philo* = love of) that take up residence in ant nests either as scavengers and nest cleaners or as purveyors of sweet honeydew for their hosts. Many beetles of the leaf litter and soil such as rove beetles, hister beetles, short-winged mold beetles, antlike stone beetles, and even tiny featherwing beetles are right at home in ant colonies. Certain mites, bristletails, some tiny cockroaches, as well as a cricket named *Myrmecophila* often steal food from their hosts or raid their refuse piles. Some ants carefully tend aphids as well as caterpillars of certain small butterflies for the honeydew that they provide. The ant hosts are so tolerant that they accept some related caterpillars that brazenly feed on the larvae of the ants. Many myrmecophiles mimic not only the chemical and tactile communications that ants use among themselves but also mimic the appearance of ants. Arthropods that share the spacious underground chambers of ants have discovered that being the guests of ants has many fringe benefits.

C. VERTEBRATES: ANIMALS WITH BACKBONES

With their backbones and internal skeletons, vertebrates are the giants of the soil. Some of them—especially the mammals—move massive amounts of earth, circulating air as well as minerals through the soil, stimulating both the growth of plants and the populations of soil invertebrates. While a few vertebrates, such as gophers and prairie dogs, feed on roots and other parts of plants, most are predators of other soil animals.

a. Vertebrates Other than Mammals

1. Salamanders

Phylum Chordata
Class Amphibia
Order Caudata
Place food web of soil: predators, diggers
Impact on gardens: absent

Mole Salamanders
Family Ambystomatidae
Size: 75–200 mm in length
Lifespan: 4–25 years
Estimated number of species: 31

Newts
Family Salamandridae
Size: 50–210 mm in length
Lifespan: 5–15 years
Estimated number of species: 74

Lungless Salamanders
Family Plethodontidae
Size: 38–200 mm in length
Lifespan: about 3 years
Estimated number of species: 376

Many salamanders are aquatic, and even the terrestrial species live only where soils are moist. Australia does not have a single salamander, and Africa only has three species. Europe and Asia have a few, but practically all the approximately 555 species of salamander, representing eight families, are found in North America (plate 37). The leaf litter and moist soils of the southern Appalachians are particularly rich in salamander species. Among these lush, green mountains, salamanders hide their lovely patterns and bright colors under rocks and in the leaf litter of the forest.

Those salamanders of the soil known as mole salamanders start out life living in ponds and breathing with gills. Later on they come ashore, losing their gills and forming lungs. Like their namesakes the moles, these salamanders stay in burrows most of the year. Although some species leave their burrows to breed in autumn, the rains of late winter and spring entice most species to migrate to nearby ponds, where they join their fellow salamanders for an evening or two of courtship, mating, and egg laying.

During their migrations, sometimes hundreds or even thousands of animals congregate in temporary ponds for this rite of spring, traversing lawns, hills, roads, or whatever else may lie between their burrows and the nearest pond. People in many communities, concerned about the threats of passing cars, have even built tunnels under roads to provide safe passage for the salamanders. Once their eggs have been deposited, the adults retire to burrows beneath the leaf litter, and every year for 10 or more years return to the same pond.

Newts are salamanders whose skins are rougher and less slimy than those of their fellow salamanders. They are found in North America, Europe, and a small portion of Asia. All newts begin life in the water, and most are primarily aquatic; but a few species spend more time on land than in water. The terrestrial phase of a newt's life is called the eft stage, and efts are noted for their bright orange and red colors. As efts slowly plod across forest floors in search of insects, woodlice, earthworms, and slugs, predators avoid these colorful amphibians with highly toxic skin secretions (plate 38).

The lungless salamanders of the soil have neither gills like tadpoles and some other salamanders nor lungs like their relatives the frogs, toads, and mole salamanders. Instead they breathe through their thin, moist skins. When examined at close range, the oxygen-carrying blood cells can easily be seen streaming through tiny blood vessels of their thin, translucent skins.

After her courtship and mating, the female goes off to a secluded spot under a log or beneath a stone, where she lays only 6 to 12 eggs and tends them for the next two months. Mother salamanders can be found coiled around their few eggs, like mother centipedes and mother earwigs that huddle with their eggs, warding off animal predators and licking off any fungi that may try to engulf their defenseless embryos (plate 39).

These lungless salamanders are found almost exclusively in North America and make up only one family of salamanders, yet they account for 376 of the 555 species of salamanders on earth.

2. Toads

The sight of a toad in the garden is a sight to gladden the heart of any gardener. Toads watch over the flowers and vegetables of a garden, alerted by any movements that might signal the presence of a slug, beetle, earwig, or other garden pest. With a flick of its long and sticky tongue, a toad gracefully snaps up these meals. Before gardens become populated with plants and pests, a toad's main concerns, however, are courting and mating. Hundreds, even thousands,

Phylum Chordata
Class Amphibia
Order Anura
Place in food web: diggers, predators
Impact on gardens: allies

True Toads
Family Bufonidae
Size: 50–90 mm in length
Lifespan: 5–10 years
Estimated number of species: 300

Spadefoot Toads
Family Pelobatidae
Size: 44–60 mm in length
Lifespan: at least 5 years
Estimated number of species: 69

of males sometimes congregate at seasonal pools in preparation for these momentous events. The musical trills of male toads fill the air over these shallow pools as the males inflate their vocal sacs, sit very erect, and look very self-important. Soon females find the performance irresistible and succumb to the charms of the males, allowing the males to embrace them as they expel their thousands of eggs in two strings of clear jelly. Within about a week, black tadpoles hatch and begin feeding on the algae in the water. Life as a tadpole is brief; the pools in which they are swimming are drying in the spring sun. In a few weeks, they sprout legs, lose their tails, and begin their overland treks to promising hunting grounds. Gardens and the night lights around our homes are often favorite destinations for toads, for they also happen to be favorite attractions for the insects that toads like to eat (plate 40).

As they set out for the first time on land, little toads face many new dangers. During their transformation from tadpole to toad, however, they picked up several new defenses that serve them well as new challenges arise. Among the many warts that they acquired when they became toads are two particularly large ones on their necks and just behind their eyes. These are the toad's parotid glands that secrete a milky white poison if the toad is provoked and threatened. A predator that tries a mouthful of toad is quickly disappointed with the flavor and often becomes very nauseated and very ill.

The arrangement of warts, spots, and splotches over the toad's skin blends in so well with the surrounding soil and leaf litter that toads can seem invisible at certain times and in certain places. Toads can actually make themselves invisible by burrowing into the soil, where they hide from heat, cold, or drought, and where they can rest during the day. Some toads even spend almost their entire lives beneath the ground, only making brief appearances aboveground to mate or to hunt on a rainy night.

Practically all toads burrow to some extent, but spadefoot toads are the ones that have special tools for digging. On the soles of their hind feet, spadefoots have horny, dark pads that expedite digging burrows that can be a few inches or several feet long. In subterranean sanctuaries of their own excavation or in a portion of some other animal's burrow that they borrow, toads can escape many of the perils of heat, cold, and predators (plate 41).

3. Caecilians

Phylum Chordata	**Impact on gardens:** absent
Class Amphibia	**Size:** 70–1,600 mm in length
Order Caecilia	**Lifespan:** 12–14 years
Place in food web: diggers, predators	**Estimated number of species:** 170

The only order of soil-dwelling amphibians besides the salamanders (order Caudata) and the toads (order Anura) is an order of totally legless creatures found only in the tropics of Asia, Africa, and Latin America. These caecilians (order Caecilia, *caecilia* = kind of lizard) resemble worm lizards and oversized earthworms whose habitats

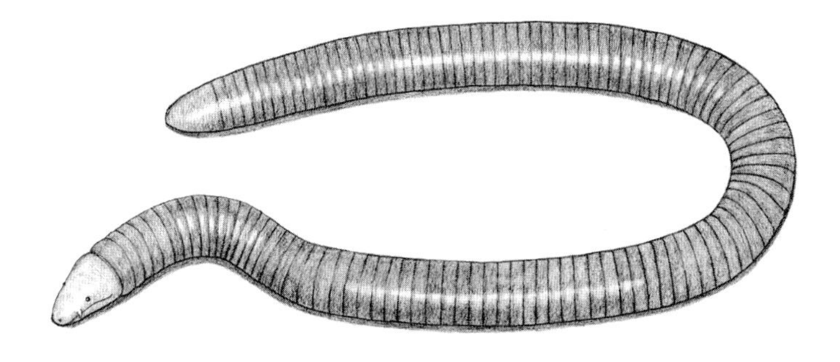

50 mm

174. Caecilians have thick, heavy skulls that they use as rams to plow through soil.

they often share. All species breathe through their moist skins and the thin linings of their mouths. In addition, although one species is lungless, most breathe with a single lung. Caecilians are burrowers with tiny eyes and a unique pair of sensory tentacles located between their eyes and nostrils. Presumably these sensory organs detect tastes and odors of earthworms and insects on which these caecilians eat. Some give birth to live young after a gestation period lasting from 7 to 10 months; others lay eggs and lie coiled around them until they hatch. Even though about 170 species have been described, we still know very little about the habits of these rarely seen amphibians (fig. 174).

4. Lizards

Phylum Chordata
Class Reptilia
Order Squamata
Place in the food web of the soil: diggers, predators
Impact on gardens: allies

True Lizards
Suborder Lacertilia
Size: 7–300 mm in length
Estimated number of species: 3,000

Worm Lizards
Suborder Amphisbaenia
Size: 180–650 mm in length
Estimated number of species: 135

Lizards of the soil are very streamlined, either having tiny legs like certain skinks and flap-footed lizards or no legs at all like legless lizards and worm lizards. As they glide through underground tunnels, lizards come across the abundance of termites, insect larvae, and earthworms that can inhabit soils having a variety of textures. For these lizards of the underground, legs would really be an encumbrance in the narrow passageways that they travel every day.

Most skinks are swift runners, as anyone who has ever tried to catch one of these sleek little lizards can attest, but sand skinks and mole skinks have legs that are too little for running. These skinks might not move swiftly aboveground, but underground they tuck their little legs against their bodies and practically swim through dry and sandy soils (plate 42).

Flap-footed lizards of Australia and New Guinea represent a family of 36 assorted species ranging from those that are burrowers and

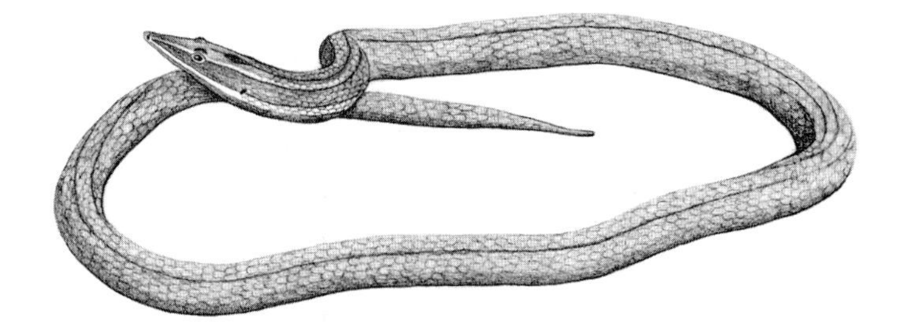

175. The elongated body of this snake lizard (family Pygopodidae) from New Guinea is an adaptation for burrowing.

resemble eyeless worm lizards to those with well-developed eyes that can easily be mistaken for snakes and that have adopted the graceful locomotion of snakes. All lizards of the family Pygopodidae (*pygo* = rump; *pod* = foot) have hind legs reduced to mere scaly flaps and have tails that are at least twice as long as their bodies. The assorted pygopodid species can be found in assorted habitats; those with well-developed eyes feed aboveground on other lizards, and those that dwell underground can always find an abundance of soil invertebrates to feed upon (fig. 175).

If it were not for its eyelids and ear openings that snakes do not have, a legless lizard could also easily pass for a snake. These lizards have very long tails that readily break off whenever they are roughed up by predators. While the rest of the lizard dashes to safety, the tail jerks and twitches long enough to distract predators that make the mistake of approaching these lizards from the rear. Snakes may look like legless lizards, but they have never mastered this trick of foiling predators by leaving their tails behind when the occasion demands a sacrifice (plate 43).

Worm lizards have neither eyelids nor ear openings like other lizards and look more like large earthworms. As the translation of their suborder name Amphisbaenia (*amphi* = double; *baen* = walk) implies, worm lizards move sometimes backward, sometimes forward; and to confuse matters even more, they have heads that look like tails. Earthworms, worm lizards, and caecilians are examples of unrelated creatures whose very similar forms have been molded by very similar environments (fig. 176).

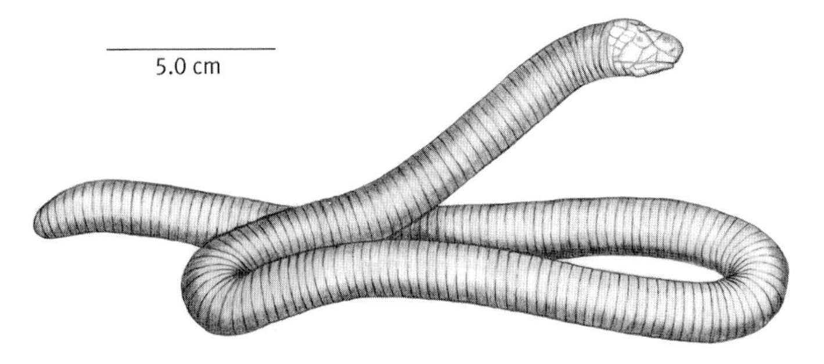

176. About 135 species of worm lizards are found in Latin America, Africa, and the Mediterranean region. One species lives in Florida.

5. Snakes

Phylum Chordata
Class Reptilia
Order Squamata
Suborder Serpentes
Place in food web: diggers, predators
Impact on gardens: allies
Size: 150–11,000 mm
Lifespan: 5–20 years
Estimated number of species: 2,700

With no legs to impede their movements in narrow underground passageways, snakes are well suited for burrowing. Not all snakes burrow, but several of them do, from large bull snakes to small garter snakes and tiny worm snakes. A snake's nose has a lot to do with its ability to burrow. Heavy nose plates apparently help snakes push and prod their way through the soil. The importance of noses in the lives of burrowing snakes can be inferred from the common names that have been given to many of them, such as hog-nosed snakes, hook-nosed snakes, long-nosed snakes, and shovel-nosed snakes. Many of these snakes, along with colorful sand snakes and poisonous coral snakes, inhabit the sands of deserts. The much damper environment of swamp mud is home to several handsome snakes with red bellies: mud snakes, swamp snakes, and rainbow snakes (fig. 177).

Larger snakes, like those from the swamps, eat rodents, frogs, toads, and crayfish; the smaller snakes of the soil, like worm snakes, blind snakes, and sand snakes, which are only about a foot long, feed mainly on earthworms and perhaps a few insect larvae.

While most burrowing snakes have noteworthy noses, one ex-

177. The patchnose snake is one of many snakes that use their noses for burrowing.

otic family of burrowing snakes from Sri Lanka and southern India, called shield-tailed snakes, is best known for its one-of-a-kind tail, which ends in a flat disc that the snake may use to stopper its burrow. These snakes of the family Uropeltidae (*uro* = tail; *pelt* = shield) rarely grow longer than 15 inches (40 centimeters) and rarely feed on anything larger than an earthworm, and they represent another example of how exquisitely adapted to the burrowing life some snakes have become.

Anyone afraid of snakes would probably change his mind after finding a ringneck snake. These gentle snakes are colorful with orange or yellow neck rings as well as bellies that are even brighter shades of orange. They are slender and rarely grow longer than two feet (60 centimeters) in length. But even they can be provoked. Rather than trying to bite with their tiny jaws, they secrete a white, stinky, and very unappetitizing fluid. Ringneck snakes like moist soils, where they live under logs, under stones, or among the fallen leaves. Here their largest prey are probably salamanders and small toads; but they will eat the slugs, earthworms, and insects that they find there as well (plate 44).

6. Turtles and Tortoises

There are several documented cases of box turtles living for more than a century. During their long lifetimes box turtles rarely venture beyond an area the size of a city block. The fortunate ones whose habitats remain undisturbed during this long period of time become well acquainted with the topography, fauna, and flora of their ter-

Phylum Chordata	**Tortoises**	**Box Turtles**
Class Reptilia	**Family Testudinidae**	**Family Emydidae**
Order Testudines	**Size:** 120–1400 mm in length	**Genus** ***Terrapene***
Place in food web: diggers, herbivores, fungivores, predators of slugs and earthworms	**Lifespan:** 50–180 years	**Size:** 115–160 mm
Impact on gardens: allies	**Number of species:** 39	**Lifespan:** 50–100 years
		Number of species: 5

ritories. They know where the earthworms are most plentiful each spring, where slugs appear on overcast days, and where mushrooms are most likely to sprout each autumn. Like all turtles of land or water, box turtles always bury their eggs in the ground. The young turtles may dig their way to the surface after about three months or may stay in the nest until the following spring. At the approach of winter or during a heat wave, box turtles retire to favorite sanctuaries in the soil and leaf litter. Part of the box turtle's secret of longevity depends not only on its ability to tightly close its shell to intruders but also on its tendency to spend some of the more vulnerable days of its life underground (plate 45).

In drier and hotter climates, larger relatives of box turtles, the tortoises, also find it expedient to hole up in a relatively cool burrow during the hottest and driest times of the year. Desert tortoises of the southwestern United States can survive miles from any body of water. Only for brief periods after a spring or summer rain do they feast on the ephemeral plants that carpet the desert floor. During the remainder of the desert year, they survive mostly on the fruits and stems of cacti. Not only can tortoises endure the desert heat, but they can also devour the spines that cover the pieces of cacti (plate 46).

Shy gopher tortoises of the southeastern United States dig large burrows 10 to 40 feet (3 to 12 meters) long that they occupy for years, returning night after night to the quiet of their refuges. Sometimes, however, these spacious refuges become crowded with uninvited guests, from small insects to large mammals like raccoons, rabbits, and possums; from harmless burrowing owls and indigo snakes to poisonous rattlesnakes and scorpions.

7. Birds

Phylum Chordata
Class Aves
Order Apterygiformes: kiwis
Order Sphenisciformes: penguins
Order Procellariiformes: seabirds (petrels, shearwaters)
Order Galliformes: pheasants, quails, turkeys, guinea fowl, chickens
Order Charadriiformes: woodcocks, puffins
Order Psittaciformes: kakapo parrots
Order Strigiformes: burrowing owls
Order Coraciiformes: kingfishers, motmots, bee-eaters
Order Passeriformes: bank swallows, thrushes and thrashers, towhees and fox sparrows, some warblers
Place in food web: diggers; predators of earthworms, arthropods, and snails; herbivores
Impact on gardens: allies
Size: 100–750 mm in length
Estimated number of species: 9,000, only a few of which dig and feed in the soil and leaf litter

While birds of most species soar through the sky and over the trees, those of a few other species delight in literally scratching out a living from the soil. Rather than catching their food while flying or in the treetops, these birds spend their days scratching about through the leaf litter and soil in eager pursuit of worms and insects. Birds alight on the ground and relate to soils in a variety of ways. Some survive on creatures that they find as they scratch through the leaf litter or poke in the ground. Others such as kingfishers, puffins, penguins, and other seabirds find their meals elsewhere but dig burrows for nesting. Finally there are those few flightless birds of New Zealand, the kiwi and the kakapo, that at all times are at home on the ground, where they scratch, poke, and also burrow.

Large birds like pheasants, quails, grouse, turkeys, and guinea fowl vigorously comb the ground and stir up whatever insects, worms, and snails lie near the surface. A number of smaller birds such as woodcocks and robins are fond of earthworms. Fox sparrows and a few warblers scratch and poke around for whatever they can find in the litter. Thrashers, thrushes, and towhees also join in the scratching when they are not searching the bushes for insects and berries (plate 47).

As long as the soil does not freeze and earthworms do not move too far beneath the soil's surface, a woodcock can continue poking about for earthworms with its long bill. Not only is the three-inch

178. Woodcocks are great fans of earthworms and frequent the same haunts in fields and open woods that earthworms do.

(75-millimeter) long bill disproportionately long for a bird that is only 11 inches (280 millimeters) long, but its legs are also disproportionately short. A woodcock's bill, however, is beautifully designed for locating and retrieving earthworms. It even has a hinge near the tip of its upper bill that allows it to open and gingerly pluck out an earthworm when the rest of the bill is buried underground. The tip of the bill is also filled with many nerves and blood vessels that endow it with fine sensitivity to odors and movements in the soil.

As a woodcock probes in the soil for food, it will hesitate a few seconds as though waiting to see if it detects any signs of an earthworm before withdrawing its bill and immediately moving on and poking again. Short legs and a short neck can actually be advantages to a bird that goes around poking its bill into the soil. A woodcock does not have far to stoop even when its bill has been poked in as far as it will go. Its large, dark eyes appear to be perched too far back on its head; but from their vantage point these eyes can survey a 360-degree field of vision. It can still keep a lookout for danger from above and from all sides while it is busily engaged in hunting earthworms. Only predators with the sharpest vision, however, have a chance of spotting this bird whose plumage blends in so perfectly with the fallen leaves among which it moves. Although most of a woodcock's attributes may seem very strange to us, they are ideal for a creature that lives among leaf litter and survives on a steady diet of earthworms (fig. 178).

Each spring and each fall fox sparrows pass through the American forests, where they tarry for about five or six weeks on their way to and from the north woods of Canada and the forests of the south-

179. During its fall migration, a fox sparrow searches for insects among fallen oak leaves.

ern United States. Often, the sound of loud rustling in the leaf litter at these times of year can be traced to a fox sparrow scratching away with sheer abandon, as it kicks aside dead leaves with both feet at once. The full range of earthy colors—umber, sienna, ocher, chestnut, russet, mahogany, chocolate—can be found on the feathers of sparrows, but the richest tawny colors are found on this largest member of the sparrow family. Years ago, when ornithologists frequently examined the contents of birds' stomachs, they found that in the spring and fall fox sparrows eat large numbers of millipedes and ground beetles of the leaf litter, arthropods that would turn the stomachs of most animals. Even though millipedes and most ground beetles secrete unpleasant chemicals, these chemical defenses do not seem to deter the fox sparrows (fig. 179).

Only a few of the over 100 wood warblers of the New World spend more time on the ground than they do in the trees. At a distance these ground warblers are easily spotted, because unlike the other warblers, they are deliberate walkers, not hoppers or runners. Ovenbirds and their relatives the waterthrushes belong to the genus *Seiurus* (*sei* = wave; *ura* = tail). As their name implies, they are constantly bobbing their tails as they promenade across the forest floor, over logs and under ferns. Worm-eating warblers belong to the equally distinctive genus *Helmitheros* (*helmin* = worm; *thero* = hunt for), a name that alludes to their prowess as hunters of worms. Warblers of the treetops are known for their bright plumage, but the earthy colors of the

180. An ovenbird inspects a mossy log for hidden insects.

181. The flightless kiwis of New Zealand use their long bills to probe deep in the soil for earthworms and insect larvae.

ground warblers blend in well with the browns, grays, olives, and whites of the forest floor (fig. 180).

Many birds nest in burrows, usually of their own making. The kakapo parrot and the kiwi are two flightless birds of New Zealand that retire to the safety of their burrows after a night spent roaming the forest floor. The long, stiff bristles surrounding their bills are touch-sensitive and help guide the birds in the dark. The wings of these birds might be weak, but the strength of the birds lies in the muscular legs and large bills that they use for digging (fig. 181).

Kookaburras and kingfishers, bee-eaters and motmots belong to an order of birds with powerful bills that they use to excavate burrows that are several feet long (fig. 182). Most of these colorful birds are tropical, but a few kingfishers are at home in colder parts of the world.

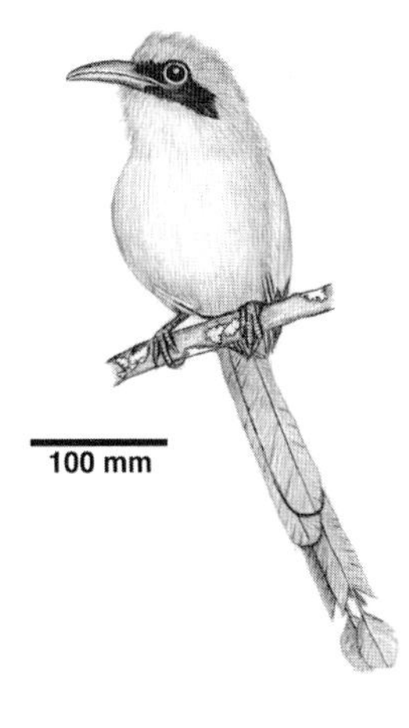

182. The colorful, elegant motmots of tropical America use their bills to dig burrows as deep as five meters.

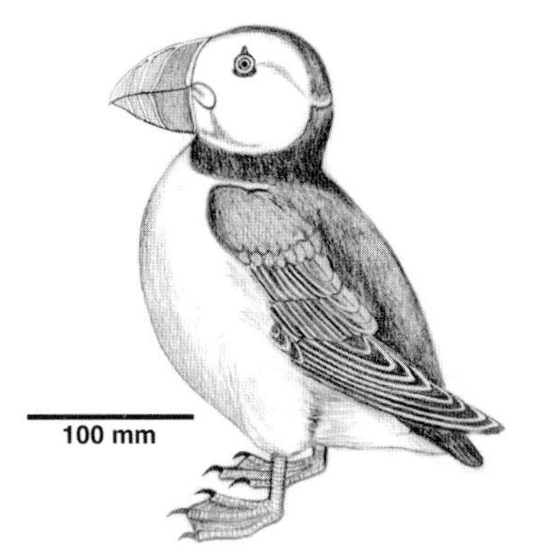

183. Puffins nest in burrows on the rocky coastlines of the Northern Hemisphere.

Puffins nest on isolated islands, where they dig holes in which each mated pair lays a single egg and vigilantly guards its precious investment (fig. 183). If danger threatens, the powerful bill and sharp claws that dug the nesting burrow can be put to use quickly and effectively. Other birds of the sea such as many penguins, petrels, and shearwaters also come ashore to dig burrows and to nest underground (plate 48).

Burrowing owls can dig their own burrows in the ground if the soil is not too hard and if no deserted dens of badgers, foxes, tortoises, or prairie dogs are available in the neighborhood. If an abandoned prairie dog hole meets its needs, a burrowing owl will move into town and will be readily accepted by the good-natured rodents. These owls even become part of the prairie dog community, helping the prairie dogs keep an eye out for intruders (plate 49).

Bank swallows and rough-winged swallows also dig their own burrows and form large colonies along stream banks (plate 50). Of all the birds he observed, the swallows were Gilbert White's favorites. In his *Natural History of Selborne*, he recorded many observations of their affairs in and around the village.

> Perseverance will accomplish anything: though at first one would be disinclined to believe that this weak bird, with her soft and tender bill and claws, should ever be able to bore the stubborn sand-bank without entirely disabling herself; yet with these feeble instruments have I seen a pair of them make great dispatch: and could remark how much they had scooped that day by the fresh sand which ran down the bank, and was of a different colour from that which lay loose and bleached in the sun.

Birds have been given the gift of flight, but some birds still choose to spend many hours of each day—or even their entire lives—scratching, poking, and also burrowing in the soil.

b. Mammals

Phylum Chordata
Class Mammalia
Place in food web: diggers, herbivores, scavengers, predators
Impact on gardens: absent (almost all species)
Estimated number of species: 4,070

We live in the Age of Mammals. When mammals inherited the earth from the dinosaurs, they quickly colonized all the land masses and all the seas. A few, such as the bats, took to the air, but most remained earthbound. Many adapted to life aboveground in trees, some to life on the ground, and others to life beneath the surface of the earth. Mammals of very different orders and families have made themselves at home on every continent and have adapted in similar ways to similar habitats.

Burrowing mammals face some of the same challenges wherever they are found and have many of the same adaptations for digging even though they are different in many other ways. Burrowing has been fashionable among the mammals. Our domesticated hamsters, gerbils, rabbits, and guinea pigs are all descended from long lines of burrowers. Spiny anteaters and duckbill platypuses, the only mammals that lay eggs, have powerful feet for digging. The spiny anteater

184. Once it decides to dig, a spiny anteater, or echidna, quickly disappears underground.

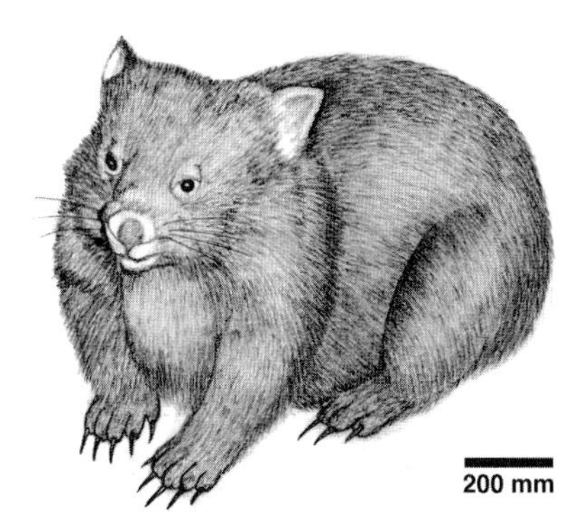

185. Wombats live in burrows that can extend up to a hundred feet.

has perfected a singular style of burrowing. Rather than digging a hole at an angle like other mammals, the spiny anteater digs a shaft straight down, soon disappearing completely from sight (fig. 184).

In addition to being home to the egg-laying mammals, Australia, Tasmania, and New Guinea are also home to the greatest diversity of marsupial (pouch-bearing) mammals on earth. Each of these marsupials has its counterpart among the mammals found in other parts of the world. To the Australians, the marsupial wombat is known as a "badger," and it is easy to see why. The wombat has short but sturdy legs and sharp claws with which it can match the badger in digging prowess (fig. 185). Bandicoots, with their long, pointed snouts

186. With its large, shovel-shaped claws, the aardvark digs a burrow about three meters long where it sleeps during the day. At other times, it uses the claws to dig into termite and ant nests.

187. The numbat or white-banded bandicoot is a squirrel-sized marsupial that uses its long tongue to lick up termites and ants.

and long foreclaws, act like skunks as they dig and poke around in the ground for insects and worms. Marsupial moles of Australia, the moles of northern Europe, North America, and Asia, and the Mediterranean mole rat of southeastern Europe all look and behave alike, even though they are not closely related. Many of the same adaptations for digging and burrowing are as likely to appear among the mammals of Australia and New Guinea as they are among the mammals of America and Europe.

Another strange group of practically toothless mammals includes the anteaters and armadillos of the Americas and the aardvarks of Africa (fig. 186). All are great lovers of termites, and their long, sticky tongues can sweep up hundreds with each lick. These animals have only small, peglike teeth that are worthless for defense, but they have stout claws that make them renowned as diggers.

Aardvarks, anteaters, and armadillos have marsupial counterparts in Australia known as numbats, or banded anteaters, that look like

188. On his visit to the pampas of South America, Charles Darwin noted that "in the evening the viscachas come out [of their burrows] in numbers, and there quietly sit on their haunches."

squirrels with digging claws, pointed snouts, and transverse stripes across their rumps. Marsupials have more teeth than other mammals, but the numbat rarely uses its many teeth. Instead, it relies on a long, sticky tongue to lick up termites and ants that it simply swallows whole (fig. 187).

Most of the world's burrowing rodents live in South America, and they occupy niches similar to those filled by woodchucks, gophers, and prairie dogs of North America. *Dinomys* (*dino* = thunder; *mys* = mouse) looks a little like a woodchuck and lives high in the Andes. Viscachas from the pampas of South America are root-eating rodents that live in community burrows appropriately called "viscacheras," their version of a prairie dog town (fig. 188). Familiar guinea pigs and rabbitlike agoutis are rodents without tails that spend the daylight hours in their underground burrows, coming out at nightfall to feed on leaves and fallen fruit.

Like the gophers and prairie dogs of the North American prairies, the South American rodents, *Ctenomys* (*cteno* = comb; *mys* = mouse) and octodontids (*octo* = eight; *odon* = tooth) are great movers and mixers of soil. *Ctenomys* has large comblike bristles on its hind feet that it uses to groom soil from its coat as well as tiny ears like the gophers of North America. It is known locally as tuco-tuco from the sound it makes in its burrow (fig. 189). The name octodontid for degus and their burrowing relatives refers to the figure-eight pattern that appears on the worn enamel surfaces of their teeth (fig. 190). Both tuco-tucos and degus are credited with helping to transform barren ground into fertile landscapes.

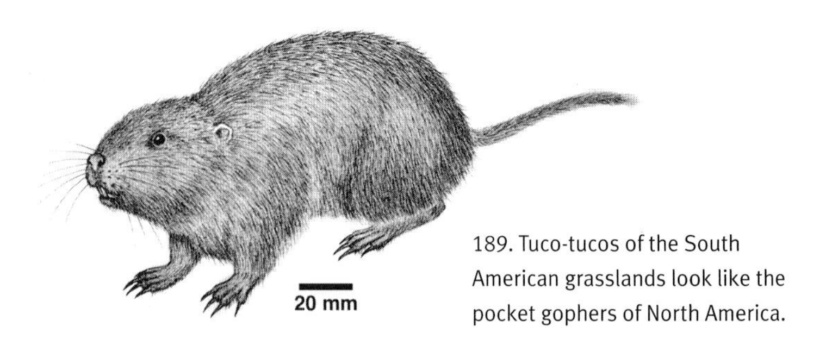

189. Tuco-tucos of the South American grasslands look like the pocket gophers of North America.

190. Degus are members of the rodent family Octodontidae that live in colonies with extensive systems of burrows and surface paths.

191. Members of a meerkat family stand tall as they survey the territory around their den.

Burrows can create a sense of community. Prairie dogs actually build underground cities with miles of tunnels extending from one entrance hole to another. The mountain beaver, or sewellel, on the west coast of the United States lives in underground colonies with extensive subway systems that lead to its favorite feeding stations aboveground. The rabbits of Europe, unlike the cottontails and jackrabbits of the North America, live in communal burrows known as warrens. On the arid plains of southern Africa, up to forty of those endearing and social members of the mongoose family, the meerkats, can share a burrow (fig. 191). But even a burrow occupied by only a single mammal can provide homes for many other animals in the neighborhood long after the original occupant of the burrow has left.

1. Woodchucks and Skunks

Phylum Chordata
Class Mammalia

Woodchucks
Order Rodentia
Family Sciuridae
Genus *Marmota*
Place in food web: diggers, herbivores
Size: 400–510 mm, head and body
Lifespan: 9 years
Gestation: 30 days
Number of species: 15

Skunks
Order Carnivora
Family Mustelidae
Place in food web: diggers, predators
Size: 330–460 mm, head and body
Lifespan: up to 10 years
Gestation: 42–66 days
Number of species: 13

From the entrance of an abandoned woodchuck den where it has lived since last fall, a skunk peeks out and sniffs the air. Skunks rarely get involved in large-scale digging projects; they usually rely on woodchucks to do most of the work for them. Because of this, the well-being of skunks is intertwined with that of the woodchucks with whom they share the forests and pastures.

When a campaign to eliminate many of the woodchucks from agricultural land in New York was conducted in the first half of the twentieth century, William Hamilton at Cornell University observed that the skunk population in an area declined every time that area's

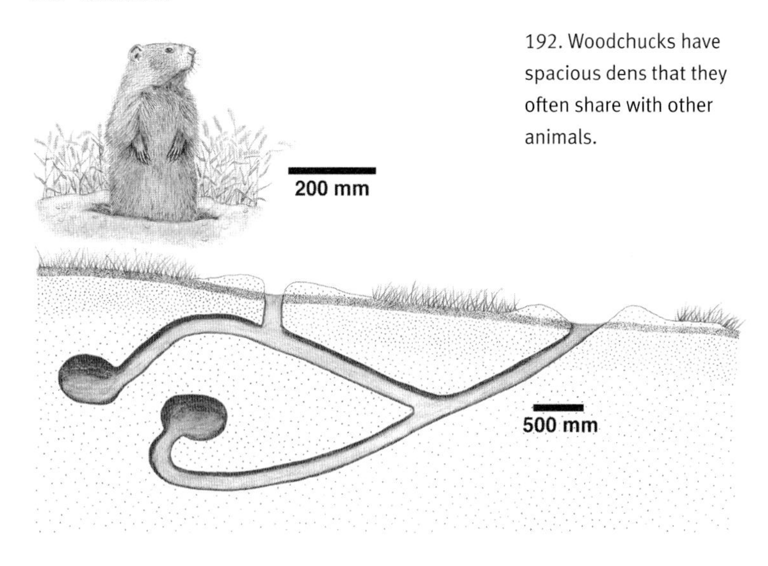

192. Woodchucks have spacious dens that they often share with other animals.

woodchucks were destroyed. Poison gas poured into woodchuck dens killed not only the woodchucks but also the skunks that had been sharing the dens. In the short-sighted act of eliminating many woodchucks to save a few alfalfa, corn, or soybean plants in each cultivated field, the exterminators also destroyed the skunks that happen to be avid eaters of beetle grubs and mice. These insects and small rodents feed in the same fields as the skunks and woodchucks, and probably do more damage to the crops than a few woodchucks could possibly do (fig. 192).

On early summer mornings you may see mother skunks and their children systematically searching for crickets, digging larvae from the soil, and inspecting cow pads for beetle grubs and other insects. Pastures and meadows are often pockmarked with their neatly dug holes, about an inch (three centimeters) wide and two or three inches (five to eight centimeters) deep. Henry David Thoreau noted the thoroughness of his neighborhood skunks. "During the succeeding half hour, it [a skunk] did not cover a space greater than three or four rods square, but literally every foot of this area was carefully inspected. Not content with rooting into every bunch of dead leaves, it dug dozens of holes, first plunging its sharp nose into the ground and then using its fore-feet, making the dirt fly." The Canadian Entomologi-

193. Skunks may not dig their own burrows, but they are constantly rooting around for insects in the soil and leaf litter.

cal Service estimated that on one eight-acre tract, skunks destroyed 14,520 beetle grubs per acre. Skunks seem to be particularly partial to these grubs and cutworm larvae. Once they discover a patch of soil grubs, they begin digging and continue to scour the area until satisfied that they have found most of the larvae (fig. 193).

Woodchucks are probably better known in certain human circles as groundhogs, the weather forecasters whose purported prowess in foretelling the arrival of spring is celebrated every year in the United States on February 2. In their own wildlife communities, woodchucks are tolerant rodents who provide new habitats and refuges for their fellow wildlife. During the winter months, while woodchucks sleep soundly in their underground chambers, other animals in addition to skunks may move in to share some of the den's extra space. Rabbits, raccoons, mice, possums, foxes, and snakes have all been known to settle down in woodchuck dens when snow falls and temperatures plummet. Abandoned woodchuck dens are often occupied as family homes by some of these same animals. Even quail, pheasants, and woodcocks may find refuge in a woodchuck den during a winter snowstorm or a prairie fire.

Clearly a woodchuck's den can be quite spacious and accommodating. While excavating a single den, a woodchuck moves about 700 pounds (320 kilograms) of subsoil along with the subsoil's associated minerals to the surface and brings air as well as organic matter to its chambers in the deeper layers of the soil. Despite their bad reputations with some farmers and gardeners, woodchucks are important and valued members in their own communities, continually improving soil conditions as well as wildlife habitat.

2. Badgers

Phylum Chordata
Class Mammalia
Order Carnivora
Family Mustelidae
Place in food web: diggers, predators
Lifespan: up to 25 years
Gestation: 3.5–12 months due to a delay in development during the winter months
Number of species: 9
Genus *Taxidea* (American badger)
Size: 420–720 mm, head and body
Genus *Meles* (Eurasian badger)
Size: 670–810 mm, head and body

Strong, muscular front legs with long claws mark badgers as powerful diggers (plate 51). Constructing new burrows, remodeling old burrows, or digging for food, badgers are often so preoccupied with their digging that they are oblivious to whomever is watching as they vigorously kick loose soil from their excavations. Even their daily toilet demands some digging, for badgers seem to be as tidy as cats, usually digging a shallow hole for their droppings and promptly covering them with soil. These badger latrines are most often located around the edges of badger territory, where they serve as territorial markers.

A badger's den is its shelter by day and a nursery in the spring. Each badger designs its den according to the terrain, soil conditions, and its individual preferences. Many badgers remodel or reuse dens that have been vacated by other badgers or other animals. The badgers of the American grasslands have a series of several dens that they dig over an area of one or two square miles (2.5 to 5 square kilometers). In this familiar home territory, a badger will never be too far from one of its dens. The badgers of Europe and northern Asia are less nomadic and more sociable than their North American relatives and settle down in dens known as setts that are often occupied by generations of badgers. Some of the more ancient setts have been continually occupied for at least two hundred years. Here a clan of anywhere from a few to as many as twenty badgers may live together.

The well-worn tunnels are often packed hard by years of badger traffic. The pounding of their paws on the hard soil below can sometimes be heard by listening at one of the entrances to the den. Setts vary greatly in their architecture, but a typical sett is a complex of interconnected tunnels and chambers that can be several stories deep, hundreds of yards in length, and can have as many as forty entrances. Its many cozy chambers are lined with grasses and leaves that the badgers use for bedding. To dig such a series of tunnels about 25 tons of soil is moved to the surface.

Hunting for daily meals demands a great deal of digging for a badger. Its favorite foods are usually hidden in the soil. The American badger is a carnivore that relishes prairie dogs, ground squirrels, and other rodents that share the prairie with it. Insects also can make up a portion of its diet. The sociable European badger has more eclectic tastes than its American relative. Earthworms are without doubt its favorite food; but depending on the season, insects of the soil, mice, rabbits, fruits, and seeds are also consumed.

Recently researchers in Britain were puzzled to find that some badger clans covered immense territories while other clans of the same size covered much smaller territories. What they eventually discovered was that the number of badgers on a given territory was related to the earthworm population of that territory. Thus, richer soils with larger earthworm populations support larger clans of badgers.

In many cultures badgers have been associated with medicine and curative powers. To many Native Americans the badger was known as the keeper of the medicine roots. Even across the Atlantic in Italy and Belgium, a locket of badger hair was believed to offer protection from sickness and evil spirits. Imagining how these beliefs arose is not difficult for anyone who has ever watched a badger in action. With all the digging that a badger does, every now and then it is bound to kick up a number of plant roots with medicinal value. The Native Americans who were adventurous enough to sample some of the roots uncovered by badgers not only were spared the effort of digging in the hard prairie soils but were also sometimes introduced to roots with remarkable healing properties that the badger had inadvertently tossed aside during its fervent digging (fig. 194).

All badgers have reputations as very able diggers; but, in addition, each of the species of badgers scattered around the world has its own noteworthy idiosyncrasies. The honey badger (*Mellivora, melli* = honey; *vora* = to eat) of Africa and southern Asia is fond of honey and young bees and has developed a collaboration with birds known as honey-guides, which are also very fond of the contents of honey bees' nests and are constantly scouting for these nests. Whenever a honey-guide comes across a nest, the bird begins a series of high-

194. Badgers are the most accomplished diggers in the mammal family Mustelidae, which includes minks, weasels, otters, skunks, and wolverines.

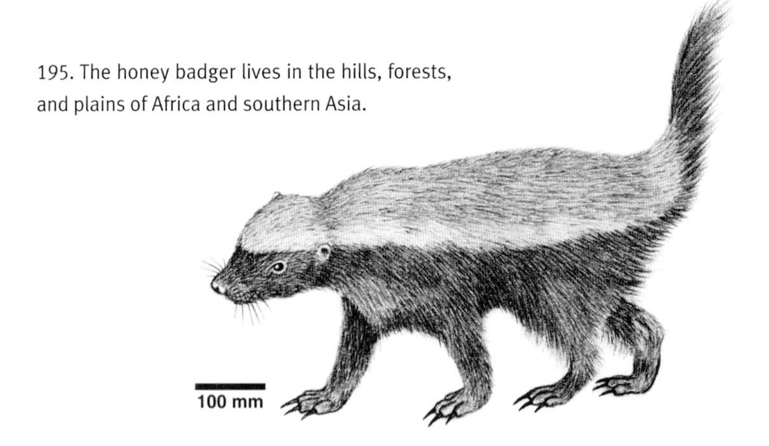

195. The honey badger lives in the hills, forests, and plains of Africa and southern Asia.

pitched calls to alert a nearby honey badger about its find. The badger follows the calls of the bird to the nest and begins tearing apart the nest to reach the honey and young bees while the honey-guide patiently waits for its share of the feast. When honey and bees are not available, the honey badger settles for meals of small animals or the contents of ant mounds and termite mounds (fig. 195).

The stink badgers of Indonesia use a stinky defense similar to that of skunks and even have the same white stripe down their backs. Any attacker that ignores this white warning stripe may have the contents of the badger's stink glands sprayed in its face.

The hog badger of southeast Asia has a particularly long snout for a badger and uses it to root around in the leaf litter and soil for earthworms, insects, and other treats. No matter where these badgers live, they all have the unmistakable physique of badgers: short legs with long claws and a short, thick neck on a stout, muscular, wedge-shaped body.

3. Prairie Dogs

Phylum Chordata	**Place in food web:** diggers, herbivores
Class Mammalia	**Size:** 280–320 mm, head and body
Order Rodentia	**Lifespan:** 8–10 years
Family Sciuridae	**Gestation:** 28–32 days
Genus *Cynomys* (*cyno* = dog; *mys* = mouse)	**Number of species:** 5

Early in the twentieth century, the artist and naturalist Ernest Thompson Seton wrote and illustrated many popular books about the animals of prairies and forests. Seton portrayed animals as intimate friends, for he had gathered his knowledge of America's wildlife from his firsthand encounters in the field, having grown up on the Canadian prairies and traveled to many of America's wild places during his long, adventurous life.

Seton had known prairie dogs since his childhood on the prairies of Manitoba. Prairie dogs were still abundant at the beginning of the twentieth century, many years after the first Europeans had settled the prairie and begun claiming land that these rodents had inhabited for many generations. Seton estimated that as many as five billion prairie dogs lived on the prairies of North America in the days before they shared the prairie with humans. Some of their towns covered up to 100 acres (fig. 196). However, human settlers were not willing to share the prairie with such a large number of rodents that littered the land with their deep holes and piles of soil.

People might not enjoy the company of prairie dogs, but prairie dogs clearly enjoy each other's company; their burrows can be located anywhere from a few feet to 50 feet (1 to 15 meters) apart. Having neighbors who can sound an alarm if an enemy like an eagle or a badger should show up in town makes life more secure on the flat and treeless prairie. Also, having a sturdy dike around the entrance

196. Prairie dogs are clearly congenial mammals that peacefully coexist in their densely populated communities.

to each burrow keeps water from rushing into a prairie dog's burrow after a heavy downpour. Since the burrows are really the only places where prairie dogs can hide on the open prairie, these rodents devote much of their time to building and maintaining their homes for protection from predators and weather. With their dexterous paws and blunt snouts, prairie dogs continually mold and pound the soil around the entrances. The mounded soil around the entrance of every prairie dog burrow acts both as a dike to keep out water after a rain and as a barricade against predators.

Being perennially curious, a prairie dog finds it hard to resist peering over the edge of its dike even when danger threatens. With its high-set eyes, a prairie dog can peer out with little more than its eyes visible above the dike. However, a threat can send the prairie dog plunging down the deep, vertical shaft of its burrow, which may be as shallow as 3 feet (1 meter) to as deep as 16 feet (5 meters). But when curiosity once again overcomes a prairie dog's terror, the prairie dog may stop at or return to a ledge or small room called the listening post on one side of the shaft just below the entrance. Here the prairie dog will wait and listen to decide whether it is safe to return to business as usual outdoors or whether it should remain hiding indoors, where it can curl up and fall asleep in its cozy bedroom.

After decades of displacement and persecution, only a tiny fraction of America's original population of prairie dogs survived. Their best defenses were no match for the deadly strychnine that humans used to poison them. Few people had anything good to say about these rodents that dug burrows and piled dirt and rocks on land that people could plow or use for grazing livestock.

To counter this negative reputation, two inquisitive scientists set out in 1947 to calculate the positive contribution that prairie dogs make to the enrichment of prairie soil. On a one-acre patch of shortgrass prairie in eastern Colorado, they counted fifty fresh mounds. The mounds of soil that prairie dogs pile around each of their burrows range in diameter from 2 to 18 feet (60 to 540 centimeters) and from 6 to 18 inches (15 to 45 centimeters) high. From the weight of soil in an average mound, the two scientists calculated that prairie dogs had piled about 22½ tons of soil on the surface. This soil, from as deep as 8 to 10 feet (two and a half to three meters), is rich in minerals that renew the fertility of the prairie, and the simple act of churning and loosening the soil makes it a more hospitable environment for the roots of plants.

A recent plan developed by the U. S. Forest Service noted that plant production is 24 percent higher in prairies inhabited by prairie dogs than in areas that are grazed by cattle. In the soils of the shortgrass prairies, where there are few deep roots of trees or tall grasses to help circulate the soil's minerals, prairie dogs help bring minerals to the surface from deep in the soil every time they dig a new burrow or expand an old one. This contribution of prairie dogs to the health and fertility of prairie soils had gone practically unnoticed and unappreciated by the early settlers.

4. Ground Squirrels and Chipmunks

Phylum Chordata
Class Mammalia
Order Rodentia
Family Sciuridae
Genus *Spermophilus* (ground squirrels) and **Genus *Tamias*** (chipmunks)
Place in food web: diggers, herbivores, predators of insects and gastropods

Size: 150–270 mm, head and body of ground squirrels
Size: 100–180 mm, head and body of chipmunks
Lifespan: up to 10 years
Gestation: 28 days
Number of species of ground squirrels: 36
Number of species of chipmunks: 24

According to early explorers of the American West, ground squirrels were sometimes more numerous than prairie dogs. Around the turn

197. A curious ground squirrel rushes to the entrance of its burrow where it can survey its dominion.

of the century, Ernest Thompson Seton estimated that as many as 5,000 Richardson's ground squirrels lived on a square mile of Canadian prairie (fig. 197, plate 52).

The underground chambers of ground squirrels are as elaborate and extensive as those excavated in prairie dog towns. Richardson's ground squirrels of the Great Plains of North America dig their burrows 3 inches (8 centimeters) in diameter, 3 to 6 feet (1 to 2 meters) deep, and anywhere from 12 to 48 feet (4 to 15 meters) long. Their relatives on the West Coast, the equally abundant California ground squirrels, can dig tunnels extending as far as 140 feet (43 meters) and having as many as 20 entrances. The burrows of the antelope ground squirrel lie only about a foot (30 centimeters) beneath the hot, dry surface of the desert soil; but within the horizontal tunnels that extend about 12 feet (4 meters) in length through hard-packed sand and gravel, the air is a cool retreat from the scorching temperatures above. Among the rocks and fallen trees of mountain slopes, golden-mantled ground squirrels dig a very similar system of tunnels that shelter them from the cold and snow of the western mountains.

Ground squirrels do not wander far from the security of their burrows. The security of having a deep, dark, and narrow burrow nearby probably gives a ground squirrel the courage to be notoriously curious about the world that is centered around its home. Ground squirrels are constantly rearing up on their hind legs with backs straight, holding their front legs close to their bellies, and bracing their bodies from the rear with their long tails as they strain to see what is happening in the neighborhood of their burrows.

Ground squirrels can be as large as tree squirrels and as small as chipmunks; they can be found in the mountains, in the deserts, and on the prairies. Judging from the large numbers of different ground squirrels that live in these places, these mammals have adapted well to life in soils with many different textures.

Most species of ground squirrels belong to the genus *Spermophilus,* a name that translates into "lover of seeds"; and several more make up the related genera of "lovers of seeds with prominent ears" (*Otospermophilus*), "lovers of seeds and sand" (*Ammospermophilus*), and "beautiful lovers of seeds" (*Callospermophilus*). In North America these different species of ground squirrels range from Mexico to the Arctic and throughout the western half of the continent; but they are absent in the eastern parts of Canada and the United States, where forests and hills replace stretches of desert and grassland. In the rest of the world, ground squirrels are found throughout most of Africa as well as across the grasslands of central Asia and a small adjoining portion of eastern Europe.

In their cozy and spacious burrows, ground squirrels and chipmunks spend the colder months of each year in hibernation and store piles of seeds that they can eat when they awaken in spring, before other seeds, green sprouts, or many insects are available as food. Because most species of ground squirrels spend so many months in hibernation, they only have time to raise one litter during the few months that they wander beyond their burrows. As though to make up for the brief time they have to court and mate, the litters that they do raise can be as large as 14 young. During these warmer months spent aboveground, they fatten up on greens and the high-protein meat of insects. In anticipation of shorter and colder days, the ground squirrels stuff their capacious cheek pouches with various seeds that they store in their underground chambers.

A ground squirrel's fondness for insects, and in particular grasshoppers, can have a significant beneficial impact on nearby crops. Much like the contributions of the much-maligned prairie dogs, however, the contributions of ground squirrels to pest control, soil aeration, and soil enrichment are rarely, if ever, acknowledged by the people who share the same piece of land. In this biased view of the deeds of ground squirrels, their occasional trespassing and destruction in cultivated fields is neither overlooked nor forgiven.

Every autumn the woods of North America, northern Europe,

and northern Asia are filled with the excited chips and chirps and chucks of chipmunks as they rustle through the fallen leaves, gathering acorns, mushrooms, buds, and many seeds. *Tamias,* the Greek name for chipmunk, means "storer," and the harvest from the forest floor will provision the storage bins that surround each chipmunk's cozy sleeping chamber. In many woods, the density of chipmunks can be surprisingly high, and chipmunk dens can be as numerous as two or three to an acre. The soil that a chipmunk moves as it digs its burrow is eventually carried off and scattered, leaving their whereabouts on the forest floor well hidden. The entrance, where most of the soil is removed, serves as a work entrance that is later plugged, while the entrance that is actually used by the chipmunk is well hidden (plate 53).

5. Moles

Phylum Chordata
Class Mammalia
Order Insectivora
Family Talpidae
Place in food web: diggers; predators of earthworms, arthropods, and gastropods

Size: 115–165 mm, head and body
Lifespan: 3–5 years
Gestation: 30–42 days
Number of species: 29

No one would consider a mole particularly handsome with its stubby legs, squat body, and beady eyes. About the only handsome feature of a mole is its soft, velvety fur, which remains spotless even in the dirtiest of the mole's tunnels. The massive digging muscles of its front legs and shoulders take up so much space at the front end of the mole's body that little space remains for a neck. As it lives out its days in dark passageways, a mole is far too busy digging and devouring earthworms and insects to be concerned about appearances. A mole has neither the eyes nor enough light in its dark galleries to appreciate the beauty of its fellow moles (fig. 198).

Highly developed senses of smell and touch are far more useful to it than keen vision. One mole, known as the star-nosed mole, has one of the most unforgettable noses in the animal kingdom. As its name implies, its nose looks just like a star, with 22 wriggling rays that look like tentacles radiating from the two nostrils. A nose like this can pick

198. Moles do most of their foraging for earthworms and insects in shallow tunnels that lie just beneath the soil's surface; they retreat to their deep tunnels where they nest and rest during droughts and the months of winter.

up not only the slightest odor but also the slightest touch. Anyone who has tried to sneak up on a mole knows how sensitive it can be to the slightest vibration of the ground. The rigid sensory whiskers or vibrissae that line the head, the tail, and the feet of every mole can amplify even the most insignificant tremor of the soil and send the mole dashing to the safety of a deep tunnel (plate 54).

Just as squirrels and chipmunks store nuts and acorns, the mole stores little balls of earthworms. Rather than immediately eating every worm that it catches, a mole will nibble off a few segments from the end of a worm and then roll the rest into a ball that it tucks into one of the crevices or cavities lining its gallery. Even though each worm in the ball is still alive, it stays paralyzed and immobilized in the mole's pantry until the mole decides to eat this particular leftover from an earlier meal.

Once its tunnels are dug, a mole regularly patrols each of them. Earthworms and larvae of insects that happen to stumble into one of these many tunnels as they roam through the topsoil may soon end up in the mole's stomach or the mole's pantry. Considering that a mole eats about half its weight each day, it must consume many generous servings of worms in its short lifetime.

Every mole excavates a series of surface tunnels for feeding and a series of deeper tunnels about two feet (60 centimeters) below, where it can nest and avoid predators as well as heat and cold. Even in compact soil a mole can progress at the rate of 12 to 15 feet (360 to 450 centimeters) an hour. As it digs, the mole shoves soil to its rear; when a sufficient amount of soil has built up behind it, the mole flips around and begins shoving the pile of soil through the tunnel and back toward the surface, where the soil eventually pours forth as a mole hill.

All this exertion and labor demands a lot of energy, as well as the food and oxygen to generate this energy. In addition to a mole's managing to find half its weight of worms in a single day, it is also able to extract oxygen from the stale air of its stuffy, poorly ventilated tunnels. A mole's large lungs make up 20 percent of its body weight. In addition, its volume of blood and the amount of hemoglobin that carries oxygen around its body are twice that of other animals its size. The demands of its subterranean life have molded the mole into a creature beautifully adapted to its poorly lit, poorly ventilated, and usually damp home.

In Africa, other animals (whose common names include "mole") occupy the digging niches filled elsewhere by moles. The molehills that punctuate African landscapes are the handiwork of either golden moles or mole rats and represent as much as half a ton of soil moved each month by a single animal. Golden moles belong to a family related to the true moles. Their name comes from the iridescent glow of their thick fur, and they have the muscular shoulders and front legs that make both true moles and golden moles such excellent diggers. Also like true moles, golden moles are insectivores and feed mostly on earthworms, insects, spiders, and snails. Mole rats, however, are rodents that survive on roots and tubers. The two front teeth, or incisors, that protrude from the mouth of a mole rat like buck teeth supply its digging power. Whenever mole rats are digging with their two front teeth, lip folds behind the incisors cover the mouth and keep the mole rats from swallowing dirt as they dig.

Of all the mole rats, only the naked mole rats form colonies in which they actually collaborate on their digging projects: one mole rat excavates with its teeth and passes soil down its line of companions until it is eventually tossed out of the molehill (plate 55).

6. Shrews

Phylum Chordata
Class Mammalia
Order Insectivora
Family Soricidae
Place in food web: diggers; predators of earthworms, arthropods, snails, and slugs

Size: 35–290 mm, head and body
Lifespan: 12–20 months
Gestation: 13–24 days
Number of species: 246

Shrews are like miniature moles that spend more time hunting aboveground than they spend digging belowground. The smallest living land mammal happens to be the pygmy shrew; a really large one can be almost two inches (five centimeters) from the tip of its nose to the base of its tail and can weigh as much as an American dime. Shrews are the busybodies of the forest floor, constantly searching for food to maintain their extravagant metabolic rate. A shrew's heart beats about 160 times a minute. Its breathing rate is also around 160 times per minute, a rate that is ten times that of most humans. A shrew's digestion is so rapid that food is converted to droppings in three hours. Living life at such a pace quickly takes its toll. For most shrews, old age arrives in about 18 months. By this time a shrew's teeth are usually worn to the gums, and it soon collapses from starvation.

In the fall and winter, you often find old, dead shrews lying among the grasses of meadows or the litter on the forest floor. Dogs, foxes, raccoons, and other animals may sniff at dead shrews but show little interest in eating them. Even possums, which have reputations for eating just about anything, do not eat shrews. Perhaps to compensate for the vulnerability of being so small, shrews emit a foul odor from glands in their skin that repels other mammals; however, the odor does not seem to discourage carrion beetles (see fig. 120). In the spring and summer, when these beetles are about, they quickly lay claim to carcasses of shrews as nurseries for their larvae.

When shrews finally take a break from hunting insects, earthworms, and snails, they retire to their tunnels under the forest litter. Here, in a nursery chamber lined with grass or leaves, each shrew begins its life. A shrew's home life can be as demanding as its frenetic outdoor activities. One female shrew can have up to 10 litters in a

199. On the forest floor, a tenrec meets a crane fly. This relative of shrews may look like a shrew, but some of the other species of tenrecs on Madagascar look like otters, moles, and hedgehogs.

200. A solenodon encounters a defensive ground beetle.

single year. Digging tunnels also takes energy, and most shrews prefer to dig their own and then fervently defend them if another shrew should have the audacity to trespass. Only one species, the least shrew, has been observed to collaborate with its fellow shrews in constructing colonial tunnels.

Shrews are found on every continent except Australia. With the exception of moles and the endangered solenodons, other members of the order Insectivora are all found in the Old World. About 12 species of hedgehogs are scattered across Europe, Africa, and Asia (plate 56); and about 30 species of tenrecs (fig. 199) are found on the island of Madagascar. Only two species of solenodons tenuously survive in the remote forests of Cuba and Hispaniola (fig. 200). Like the shrews and moles, most of these other insectivores are diggers, and they all seem very fond of earthworms and other soil invertebrates. Their long, probing snouts have noses with acute senses of smell and long whiskers with fine senses of touch. Even in the disarray of leaf litter and the dark labyrinths of the soil, these snouts can track down the most elusive insects and worms.

7. Pocket Gophers

Phylum Chordata	**Size:** 130–225 mm, head and body
Class Mammalia	**Lifespan:** 4 years
Order Rodentia	**Gestation:** 17–20 days
Family Geomyidae	**Number of species:** 34
Place in food web: diggers, herbivores	

On the vast and fertile grasslands of the world—the prairies of North America, the steppes of Asia, and the savannahs of Africa—herds of grazing animals have dined and left their droppings generation after generation, enriching the soil but at the same time pressing and packing the rich soil beneath their hooves. The plants that live on this soil are particular not only about the nutrients and water that the soil contains but also about the ease with which their roots can grow and breathe in the soil. In the economy of the grasslands, it is the diggers that continually are restoring the spongy, crumbly structure to the soil that the hoofed animals continually compress and compact. With their incessant urge to burrow, the rodents, badgers, and insects of the grasslands are ceaselessly mixing layers of soil and re-creating the rich, porous structure of soils in which plant roots grow so well.

While most burrowing animals such as prairie dogs, badgers, and woodchucks spend a good portion of their waking hours traveling aboveground, the gophers of the grasslands almost never leave their burrows. Here in their subterranean tunnels all their needs are met: protection from owls, hawks, and other predators, shelter from bad weather, as well as a bountiful supply of roots and tubers. Whenever traveling to new destinations, gophers simply dig their way there. A lifetime of traveling underground can add up to a lot of digging and a phenomenal amount of earth moving. A gopher can dig 300 feet (100 meters) of tunnel in a single night. If the ground gets hard and the going gets tough, the gopher starts using its teeth as well as its claws.

Even if gopher burrows only occupy one tenth of one percent of a soil's surface, each year on every acre of land 250 pounds (115 kilograms) of soil will be moved from the subsoil to the soil's surface,

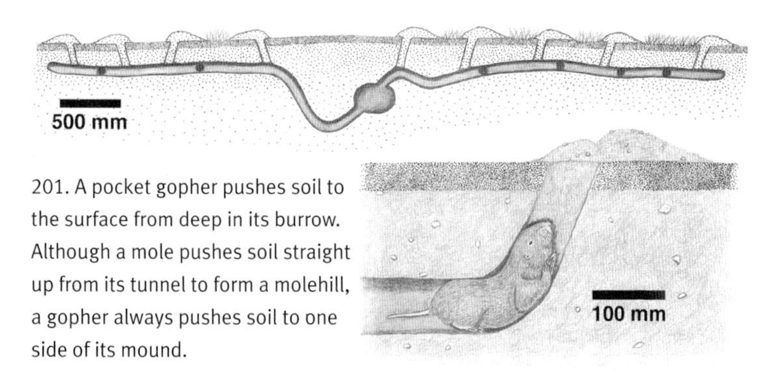

201. A pocket gopher pushes soil to the surface from deep in its burrow. Although a mole pushes soil straight up from its tunnel to form a molehill, a gopher always pushes soil to one side of its mound.

where new rocks will be weathered and new minerals will be freed for growth of plants. Gophers can be so abundant in places that their new mounds may cover 15 to 20 percent of the soil's surface and involve the movement of many tons of soil in one year alone. One naturalist estimated that pocket gophers move 8,000 tons of soil each year in California's Yosemite National Park (fig. 201). These are impressive figures for a small mammal that is less than a foot (30 centimeters) in length from the tip of its nose to the end of its short, naked tail. For a life of constant digging in dark, narrow tunnels, gophers have just the right combination of traits—tiny ears for tight tunnels, tiny eyes for dark tunnels, and strong front legs for digging. Numerous calculations have been made of how much soil is moved by each rodent as well as how abundant each one is, and gophers always seem to score highest on all counts.

The name gopher can be traced to the French word *gaufre,* meaning honeycomb. One gopher's tunnels can extend over an acre or more. The soil can be so perforated by these tunnels that a cross section of the soil inhabited by a gopher really does look like a honeycomb. The main tunnel of a pocket gopher home is about four inches (10 centimeters) in diameter and between six and nine inches (15 and 24 centimeters) belowground, and it may run for 500 feet (150 meters). Many side tunnels branch off the main tunnel, some tunnels leading to the surface, where excavated soil is pushed out of the opening and spread in a flat, fan-shaped mound. Other tunnels lead to toilet chambers, feeding chambers, and storage chambers. One tunnel descends several feet to the gopher's grass-lined nest chamber. An intricate network of twisting tunnels connects the many mounds of a pocket gopher's home.

Rarely does a pocket gopher appear aboveground. In its tunnels it can find all the roots, bulbs, and underground stems that it needs to satisfy a healthy appetite for half its weight in food each day. After each meal the gopher's droppings and scraps are left to mix with the soil in the tunnels, adding organic matter and enriching soil well below its surface. When the tunnels need expanding, the pocket gopher moves mineral soil from the depths of its burrow to the surface, where it is mixed with the organic remnants of plants. So much exchange of organic matter and mineral matter takes place around gopher tunnels that the surface of the ground in the vicinity of a gopher's den probably turns over at least once every two years. The contributions of gophers to overgrazed grassland are easily spotted from afar as oases of lush, green grass surrounded by stunted grass struggling to grow in hoof-trodden soil. Wherever gophers have turned and crumbled the soil, the plants thrive in the hospitable environment.

8. Kangaroo Rats

Phylum Chordata	**Place in food web:** diggers, herbivores
Class Mammalia	**Size:** 100–160 mm, head and body
Order Rodentia	**Lifespan:** about a year
Family Heteromyidae	**Gestation:** 33 days
Genus *Dipodomys*	**Number of species:** 22

Beneath the desert mounds that mark the homes of kangaroo rats lives a whole menagerie of desert creatures. Toads, snakes, lizards, centipedes, millipedes, ants, beetles, crickets, scorpions, and roaches are known to move in and share the spacious passageways that lie from two to three feet (60 to 90 centimeters) below the surface and that can occupy three or four levels as they twist and turn in this or that direction. The kangaroo rat seems to tolerate all of these tenants and provides them a cool refuge from the heat of the desert sun (fig. 202).

Here in its underground chambers, the temperature almost never rises above 85°F (30°C) even though the temperature at the surface of the soil may rise as high as 160°F (70°C). The kangaroo rat avoids the desert heat and many desert predators by sealing itself in its bur-

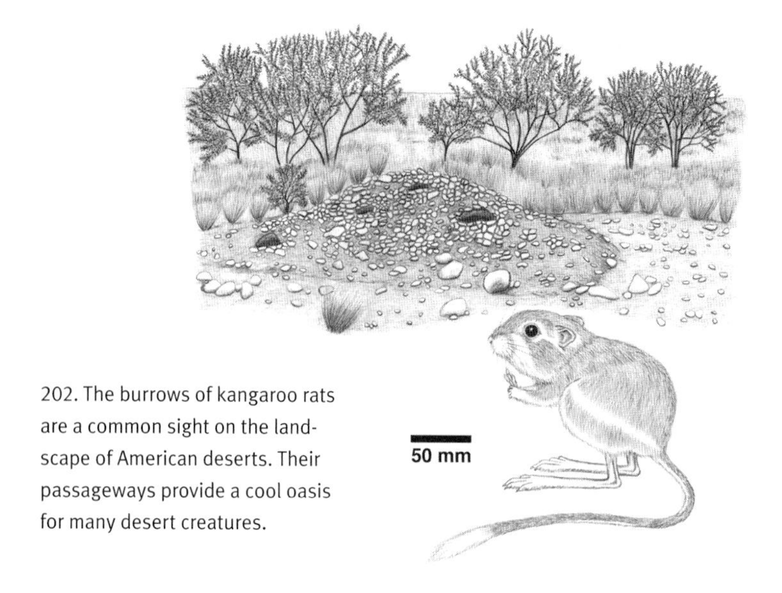

202. The burrows of kangaroo rats are a common sight on the landscape of American deserts. Their passageways provide a cool oasis for many desert creatures.

row and by resting during the daytime hours. Kangaroo rats cannot withstand much heat and soon succumb if they are exposed for only a few hours to temperatures over 100°F (40°C).

Most animals keep cool by losing heat as water evaporates from their bodies. Kangaroo rats, however, do not use this standard strategy for keeping cool. They do not pant like a dog or sweat like a horse. They instead stay cool and at the same time conserve precious water by avoiding the heat and concentrating their urine. They also lose very little water in their hard, dry droppings; and what little they do lose from their bodies, they manage to reuse. By eating their droppings, kangaroo rats manage to extract any remaining water and nutrients that their digestive tracts failed to absorb the first time around. Kangaroo rats are quintessential water conservationists, surviving on whatever water they produce from digestion of their diet of desert seeds and conserving even this small amount of water as they spend their days beneath their desert mounds in the relative coolness of their underground sanctuaries.

In some areas of the desert, the mounds of kangaroo rats cover as much as 30 percent of the land area. Although well-beaten trails lead from burrow to burrow, about the only time a kangaroo rat actually shares its burrow with another kangaroo rat is during the mating

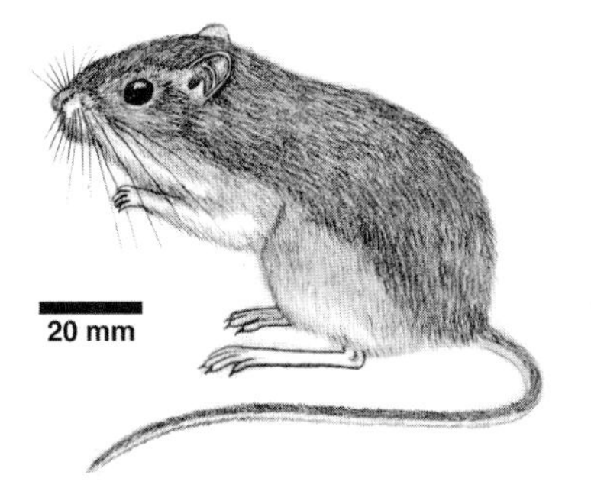

203. Pocket mice spend nights filling their fur-lined cheek “pockets” with seeds of desert plants and return to their burrows to unload the harvest.

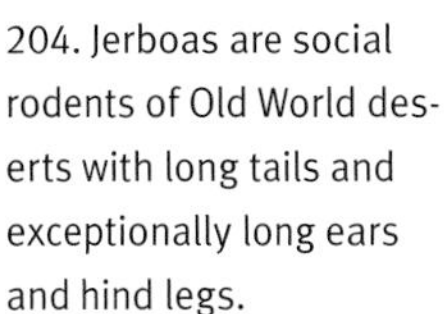

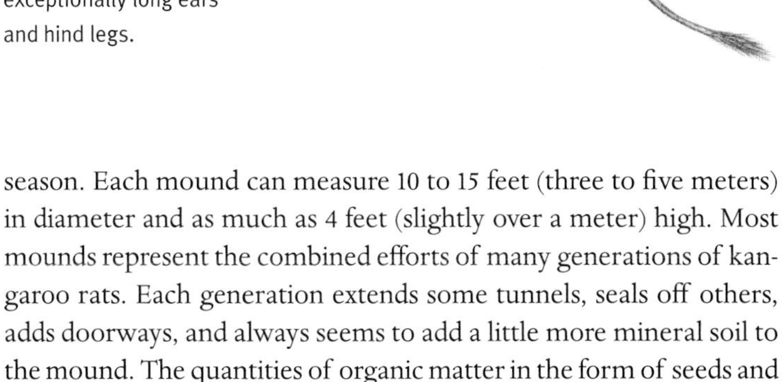

204. Jerboas are social rodents of Old World deserts with long tails and exceptionally long ears and hind legs.

season. Each mound can measure 10 to 15 feet (three to five meters) in diameter and as much as 4 feet (slightly over a meter) high. Most mounds represent the combined efforts of many generations of kangaroo rats. Each generation extends some tunnels, seals off others, adds doorways, and always seems to add a little more mineral soil to the mound. The quantities of organic matter in the form of seeds and dried grass that kangaroo rats carry into their burrows commonly range from 3 to 6 bushels; 14 bushels of plant matter was found in the burrow of one particularly industrious kangaroo rat.

A number of other small mammals seem to have much in common with these handsome, big-footed rodents. Kangaroo rats share their desert habitat with closely related, but even smaller, pocket mice (fig. 203). These dainty mice with big hind feet burrow into the desert soil and survive on the seeds that desert plants are able to provide.

On the other side of the earth, in the deserts of Asia and Africa, another kangaroo-like rodent, the jerboa, has hind feet that are even larger than those of any of its American relatives (fig. 204). What is equally striking about the jerboa are its big ears. Some species of jer-

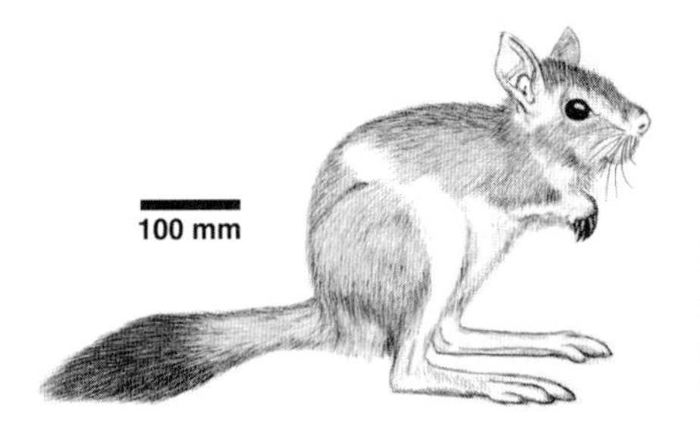

205. The springhare is a rodent that, on first encounter, could be mistaken for a kangaroo or a rabbit.

boas have ears that are half the length of their three-inch (80-millimeter) bodies. The jerboa belongs to the family Dipodidae (*di* = two; *podi* = feet) and gets around fine on its two big feet, holding its small front legs above the ground. The deserts and savannahs of Asia and Africa are also home to the familiar gerbils that could be mistaken for jerboas if their ears and hind legs were larger. In their deep burrows, insulated from heat and cold, gerbils and jerboas hoard seeds and socialize.

On the grasslands of south and east Africa live some larger big-footed burrowing rodents called springhares, which have been assigned a family of their own by biologists (fig. 205). Their family name, Pedetidae, translates to "leapers" and seems an appropriate choice of names for these rodents that can leap 10 to 13 feet (three to four meters) with their six-inch (150-millimeter) feet. With the ears as well as the head of a rabbit, the body of a kangaroo, and a tail slightly longer than its 14- to 17-inch (360- to 430-millimeter) body, the springhare looks as though it has been pieced together with parts from different animals.

Among the kangaroo-like rodents, however, the jumping mouse holds the record as the most accomplished jumper of them all, its three-inch body clearing 12 feet (three to four meters) in a single leap (plate 57). While the other leapers inhabit the deserts of the world, jumping mice live in the moist woods and fields of North America. They dig burrows in which they spend half of each year in hibernation.

PART THREE

Working in Partnership with Creatures of the Soil

Humans have transformed between one third and one half of the earth's surface with urbanization, tilling, logging, and grazing. Humans have also transformed the cycling of such chemicals as carbon dioxide, water, and nitrogen, as well as toxic metals and compounds in the atmosphere. "We are changing Earth more rapidly than we are understanding it," as four biologists wrote in a 1997 article in *Science.* It is our responsibility to manage these changes without devastating the habitats of other species that share the earth, for their sake as well as for our own.

The great issues facing our environment—both locally and globally—are linked to the innumerable organisms that live underground. The health of our soils in gardens and fields has suffered from a dependence on chemical fertilizers and a lack of appreciation for the contribution to soil fertility of the myriad creatures that labor underground. Their interactions belowground result in a balance between the processes of growth and the processes of decay. The humus that these creatures generate during the processes of decay is essential to fuel the processes of growth. Without humus, mineral nutrients from fertilizers are soon lost from the soil, along with the pore spaces that hold the moisture and the air that make a well-structured soil so productive and fertile. Countless reciprocal interactions between life belowground and life aboveground shape the world in which we live.

1. Preventing Erosion

Erosion as a natural process shaped the familiar prairies and mountains and river valleys that grace the surface of the earth. It is part of a natural cycle of wearing down and building up that has always marked the history of the earth, with the rate of soil formation keeping pace with the rate of soil erosion. Soil formation varies from place to place around the world because the rate at which a soil forms depends on several features of a place—its climate, its soil-forming rocks, its flatness or steepness, its plants and animals—but a rough estimate of the average rate of soil formation has been given as a foot of soil developing every 10,000 years

It is one of the ironies of nature that the same forces of wind and water that help create soils can also do a thorough job of destroying soils. Each year wind and water erosion are notorious for carrying away from cultivated fields not just thousands or millions of pounds, but billion of tons of soil. Prior to the arrival of humans on earth, erosion from natural forces of wind and water amounted to an average net loss of 88 feet of soil every one million years. Today, however, erosion attributable to human causes results in an average net loss of soil of 1,200 feet every million years. As a geologist at the University of Michigan calculated, soil lost at this rate would completely fill the Grand Canyon of Arizona in 50 years.

In many locations, the intervention of human farming has clearly disrupted and distorted the natural cycle of soil erosion and soil formation, accelerating soil loss to such an extent that the normal processes of soil regeneration simply cannot keep pace; on other land, judicious farming practices have actually built soil at rates faster than this estimated natural rate of soil formation.

Tillage is the way we mechanically work the soil to grow crops and to control weeds, using plow, disk, and harrow. By exposing soil to the elements of wind, rain, and snow, tillage dramatically modifies the environment of the soil and exposes it to erosion (plate 58). Far fewer plant species now dwell on the land—often only the species that is being grown as a crop along with the few weeds that managed to escape herbicide and harrow. Compared with native vegetation, crop plants leave behind far less organic matter; their root systems cover less territory underground and bring up fewer nutrients to the surface.

The conventional tillage that has been practiced since the birth of agriculture is being replaced as gentler and less disruptive methods of tilling the land are being tried and perfected. After the harvest, the crop residues that linger on the field have traditionally been plowed under, exposing the bare earth, but today many farmers choose to leave residues on the surface of the field to protect against erosion and to provide a haven for the decomposers and the makers of humus. These same farmers also leave strips of land untouched by the plow, where trees are planted as windbreaks and where fence rows are allowed to harbor a variety of plants and animals. This practice of reduced tillage, known as conservation tillage, may have the drawback of not controlling weeds as well as conventional tillage; but it does save on fuel, equipment use, and reduces erosion by 40 to 50 percent (plate 59).

At the Land Institute in Salina, Kansas, researchers are experimenting with perennial crops that would require no additional tillage after being planted. Unlike annual crops such as corn, soybeans, wheat, and rye, whose roots and shoots die back every autumn, perennial crops have root systems that survive for many years and shoots that resprout every spring. During the winter the living roots of perennial crops would hold the soil in place, and the remains of their leaves and stalks would provide both cover and nutrients for the soil. The plant debris left behind at the end of each growing season would provide a hospitable home for a large and diverse population of soil creatures. Annual autumn tillage could be abandoned, and the soil beneath the perennial crop could be spared the ravages of winter winds and spring rains.

Whenever soil blows or washes away, the nutritive elements of the soil disappear as well. What goes first is the topsoil. What remains after the topsoil has disappeared is the far less productive and less fertile subsoil. As the topsoil disappears so do the humus and plant nutrients that are concentrated in topsoil (fig. 206). Humus and organic matter hold water and mineral nutrients in the topsoil and provide a home for the microbes and animals that are responsible for the formation of humus and organic matter from plant and animal debris. Humus in cool and temperate climates decomposes slowly and releases only a small percentage of its nutrients each year. As long as they are continually replenished, organic matter and humus are a storehouse for nutrients that plants can use.

206. Dry weather and conventional tillage allow wind to erode soil. (Science VU/ARS/Visuals Unlimited)

2. Avoiding Excessive Use of Fertilizers

Nutritive elements are lost from soils not only by wind and water erosion. They can be depleted from the surface soil as rain or snow soaks deep into the soil and carries elements out of reach of most plant roots. A soil that is spongy and rich in humus, however, will bind many of the soil nutrients and prevent their being rapidly depleted from the topsoil. Earthworms, other deep-burrowing animals, and deep roots of plants help retrieve nutrients that have been leached from the topsoil and recirculate them to the soil's surface.

Elements are also lost from the soil every time crops and animals are taken from the farm to the market. As crops grow and as livestock graze on the farm, they directly or indirectly take up elements from the soil. All plant crops obtain elements from the soil, and livestock obtain elements from these crops. Some crops and farm animals remove more of certain elements than do others. Each bushel of corn and each bushel of wheat contains about a pound of nitrogen, a quarter pound of phosphorus, and a quarter pound of potassium that came from the soil in which they grew. Crops of legumes like soybeans, clover, and alfalfa certainly add nitrogen to the soil, but they also remove many pounds of calcium, phosphorus, and potassium from the soil.

A thousand-pound cow indirectly takes about 25 pounds of nitrogen and about seven pounds of phosphorus from the soil where it grazes. No wonder that each time crops and livestock are taken from farm to market, the farm's soil becomes a little poorer in nutrients. As animals and plants are harvested from the land that nourished them, all the elements they took from the soil go with them. The nutrient cycles of farm soil are restored when the nutrients used by plants and animals during life are returned to the same soil after their death.

The constant demand on farmers to increase their output of crops to feed ever-increasing populations of humans and livestock, as well as the constant demand by farmers that their land be more profitable, has established a sinister pattern of accelerated loss of nutrients from the soil and accelerated consumption of synthetic fertilizers to make up for these losses. Fifteen times as much nitrogen, four times as much phosphorus, and thirty-six times as much potassium were added to cultivated fields in 1950 than had been added fifty years earlier. In the United States we used at least three times as much fertilizer in 2005 as we used in 1975.

Synthetic nutrients are continually poured on agricultural fields in the mistaken belief that increasing the production of a soil also increases the fertility of a soil. Commercial fertilizers can quickly boost plant nutrition and the production of a soil; but soon the cations like calcium, potassium, and magnesium are leached from the topsoil and washed out of reach of plant roots unless they are bound by negatively charged particles of humus.

Without the spongy, crumbly structure that humus gives to a fertile soil, many nutrients are simply leached from the topsoil by rain before roots have a chance to put them to use in building plant tissues (plate 60). By adding too much nitrogen to the soil in the form of nitrate fertilizer, the leaching of positively charged nutrients from humus and topsoil is actually accelerated as the negatively charged nitrates attract positively charged cations and drag them deeper into the earth and eventually into groundwater, streams, and lakes. In these aquatic environments, excess nutrients from the fertilizers that have passed beyond the reach of roots stimulate growth of algae and contribute to pollution of the waters.

If this situation were not bad enough, denitrifying bacteria put the excess nitrates to use by converting them to toxic nitrites and to nitrogen oxides, which are atmospheric pollutants. One of these oxides of nitrogen, nitrous oxide (N_2O), contributes to both global warming and acid rain. Fertilizing fields with common mixes of nitrogen, phosphorus, and potassium may often aggravate deficiencies in other nutrients because excess levels of these important nutrients can actually make a soil more acidic and depress the plants' uptake of other nutrients such as calcium, magnesium, and zinc.

Soil without humus and with few living creatures will continue to produce only as long as fertilizer is added. The soil has no nutrient reserves of its own unless it also has humus to hold nutrients where plant roots can reach them.

How do we return something to impoverished soils that might make a more lasting impression on its fertility than pouring pound after pound of fertilizer on soil that has no place to store these nutrients? Compost, mulches, animal manures, and green manures not only contain the essential nutrients for plant growth but they also add large amounts of organic matter to the soil (plate 61). Addition of the organic matter improves the structure of the soil, provides a refuge for organisms that live underground, and by acting as a storehouse for nutrients, slowly releases them at rates of about 2 to 4 percent each year.

Oats, rye, mustard, and legumes are often planted at the end of summer for use as "green manures" after the primary crops of summer such as corn and soybeans have been harvested. These cover crops grow quickly, protecting the soil from the rains and winds of fall, winter, and spring. Whatever fertilizer was not taken up by the primary crop plant during the summer can now be salvaged by the fast-growing roots of the green manure before nutrients from the fertilizer are leached from the land and washed into the groundwater.

When cover crops are plowed under the following spring to make way for the primary crop, their remains make life more hospitable for creatures of the underground. These creatures can do what no human can do alone. They improve the structure of the soil by decomposing plant remains to humus and liberating the mineral nutrients that these plants recovered from the earth during their growing days.

Even the weeds in a garden or field serve a useful function as a form of green manure. Weeds have a knack for partitioning resources of the soil. By spreading their roots different distances both horizontally as well as vertically, plants can tap different mineral resources.

Some weeds have a real appetite for particular minerals and gather them up from the nearby soil. Their roots bring up minerals from belowground, and whenever the weeds are hoed or pulled their remains leave a dose of minerals and organic matter on a compost pile or on the surface of the garden's soil that is within easy reach of the roots of garden vegetables.

3. Effects of Acid Rain

Hydrogen is probably the single most important element in the soil, not because it alone is so important for the well being of plants but because it is so important in determining the availability and solubility of just about all the other essential elements of the soil. Soils with relatively high concentrations of hydrogen ions—greater than one part to 1,000,000—are acidic soils, and soils with relatively low concentrations of hydrogen ions—less than one part to 100,000,000—are alkaline soils. Soils with concentrations of hydrogen ions somewhere between these two concentrations are considered slightly acid, slightly alkaline, or neutral soils, where most plants seem to grow best.

In acidic soils some elements form insoluble compounds with other elements, while in alkaline soils another group of elements forms other insoluble compounds. Since plant roots can only use soluble elements and compounds, elements in the soil become unavailable for plant use whenever they combine with other elements to form insoluble compounds.

While acidic soils have sufficient concentrations of available iron, zinc, nitrogen, and manganese, they are deficient in soluble phosphorus, potassium, calcium, sulfur, and magnesium. This is why plants showing iron, manganese, or zinc deficiency are often treated by adding hydrogen ions to the soil rather than by adding iron, manganese, or zinc. These elements are already in the soil; they simply are not soluble and available for uptake by roots. In acidic soils with high concentrations of hydrogen, these elements become soluble and free to be taken up by roots. Too many hydrogen ions, however, can create real problems for roots and other inhabitants of the underground as we are learning from recent studies of acid rain.

Rain rapidly leaches nutrients from soil that is poor in humus, but acid rain does an even more thorough and sinister job. By contributing to the acidity of soil, acid rain can have devastating effects on life underground, setting off a chain of events that soon has dire consequences for life both above- and belowground.

Acid rain forms when moisture in clouds combines with sulfur oxides and nitrogen oxides released mainly by automobiles, factories, and power plants. Some of the nitrogen oxides also arise from excessive use of fertilizers. These nitrogen gases are released into the atmosphere when denitrifying bacteria of the soil break down nitrates from fertilizers. Sulfur oxides react with water to form sulfuric and sulfurous acids; nitrogen oxides react with water to form nitric and nitrous acids. When these acids fall to the earth, like other acids, they release hydrogen cations (plate 62).

Not only do the numbers of hydrogen cations released often exceed the numbers of other nutrient cations but they also bind more strongly than these cations to negatively charged particles like clay and humus. As hydrogen cations from acid rain percolate through the ground, they displace other nutritient cations such as calcium, magnesium, and potassium from soil particles, pushing them beyond the reach of plant roots (fig. 207).

Other cations like aluminum that are toxic to plant roots are insoluble and safely impounded in most soils—but not in acid soils. This is because anytime the hydrogen ion concentration of the soil becomes particularly high, toxic cations, such as aluminum or cadmium, are replaced by hydrogen ions from the humus and clay particles that bind them. These toxic cations are now free, soluble and harmful to plants and animals.

Along with the acid rain created from oxides of sulfur and nitrogen, winds carry the toxic element mercury from coal-fired power plants. When mercury is carried to the earth with acid rain, it is apparently transformed in the soil into a form taken up first by worms and insects and then by the birds that feed on soil creatures. Mercury in the environment is a poison that does not break down. Acid rain not only leaches essential elements from the ground but it also releases toxic elements that poison life in the soil and eventually life aboveground.

4. Avoiding Salt-Encrusted Soils

As shown in fig. 207, cations such as hydrogen, calcium, magnesium, and aluminum are strongly adsorbed to surfaces of clay and humus particles. The positive charges of these cations in turn attract and

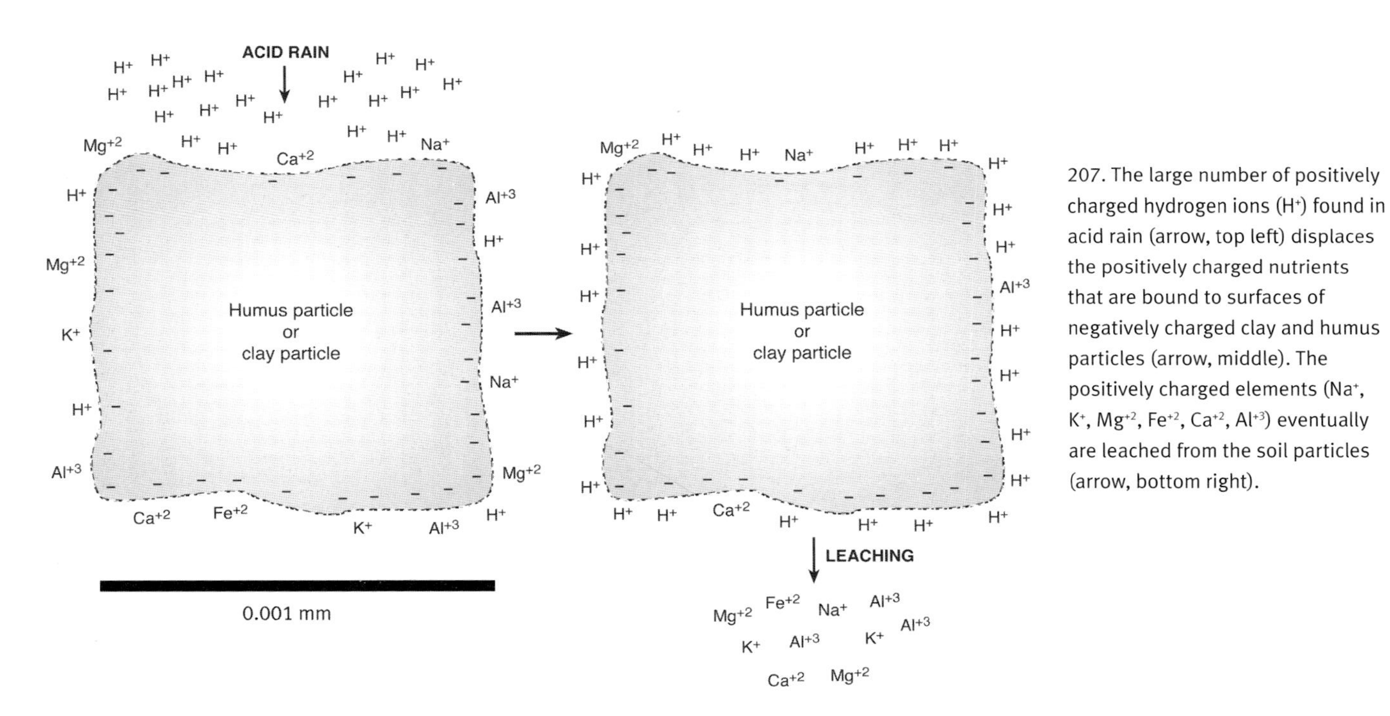

207. The large number of positively charged hydrogen ions (H^+) found in acid rain (arrow, top left) displaces the positively charged nutrients that are bound to surfaces of negatively charged clay and humus particles (arrow, middle). The positively charged elements (Na^+, K^+, Mg^{+2}, Fe^{+2}, Ca^{+2}, Al^{+3}) eventually are leached from the soil particles (arrow, bottom right).

hold together these negatively charged particles. This interaction of cations with particles of humus and clay initiates formation of soil aggregates that give soil its spongy and crumbly structure. Algae, fungi, and bacteria then produce the complex organic compounds that also hold these particles together and help stabilize these soil aggregates. Soil aggregates give soil its structure and determine the ease with which air and water as well as animals, roots, and fungi move through the soil (see fig. 19).

However, salts of sodium, potassium, calcium, and magnesium can naturally accumulate to such high levels that they form white crusts on soil surfaces in arid regions wherever rain and snowfall are not sufficient to flush these salts into deeper soil layers (plate 63). Such soils are often poorly drained, and the water table lies not far underground. Vegetation in these regions normally draws up the water through its roots, helping to keep the salts and the water level in the lower layers of the soil. But if this vegetation is removed from the soil or if these arid soils are irrigated with salty water, the salts that started to leach to lower soil layers are brought to the surface again by evaporation and capillary action, forming a white, salty crust.

When salts accumulate to such high levels in soil, sodium cations, in particular, disrupt soil structure, impede infiltration of water, and clearly interfere with plant growth. Of all the salts, sodium adsorbs most weakly to soil particles and fails to prevent the negatively charged particles of humus and clay from repelling each other and dispersing. The dispersed particles of humus and clay then clog the pores that once existed in the soil and obstruct the flow of air, water, and nutrients. Not only do salts hold water and make this water unavailable to plant roots, but sodium salts also break down the aggregates of crumbs and granules that give healthy soils their spongy structure.

5. Maintaining Soil Structure

Healthy soils with humus and organic matter support an abundance of animal and plant life. The structure of these soils absorbs water just as a sponge does, and the water they hold measures about six times their weight of humus and organic matter. Rain that falls and snow that melts on these soils quickly infiltrates and soaks the spongy soil rather than rushing downhill and carrying soil particles with it. The water travels throughout the ground eventually to emerge in clear bubbling springs. Where spongy soil covers the hills above a stream, the waters flow clear and unhurried after most rains.

Soils, however, lose their spongy structure through certain human activities, shedding their nutrients and water like the back of a duck. Improper irrigation and loss of vegetation on arid ground often contribute to the formation of salt-encrusted soils and the breakdown of soil structure. Land that is low in organic matter, that is over-tilled, or that has been compacted by equipment or livestock has soil that has lost its vitality. Clear-cutting of forests by loggers and the practice of slash-and-burn agriculture in the tropics expose soil to compaction and loss of nutrients from organic matter. Constant scavenging of the earth's surface for firewood and animal dung destroys soil structure and deprives soil of important nutrients (plates 64–65).

Any slope to the abused land sends water rushing downhill to nearby streams rather than quickly absorbing the water among its rocks, mineral particles, and organic particles, where it can be slowly released later. The more water that stays on the surface of the soil and the faster it flows, the more soil it carries along with it. Once the rich life of the soil and the humus it produces are lost, the spongy structure of a soil disappears, and the streams that drain this soil now rush by muddy and overflowing after a summer rain or a spring thaw. The rushing streams carry nutrients from the land, diminishing its fertility after every rain and snow melt.

Great civilizations of the ancient world arose in river valleys, where fertility of the land was replenished each year by natural floods. The civilizations that arose along the Tigris and Euphrates, the Indus, the Mississippi, the Nile, and the Yangtze endured as long as their soils remained fertile.

The grandeur of ancient Egypt was based on gifts from the Nile. Each spring, as the rains fell across northern Africa, humus and clay particles from the rich soils of the upper Nile were carried to Egypt in the floodwaters of the river. Accompanying these negatively charged particles was a rich supply of positively charged nutritive elements such as potassium, calcium, and magnesium, as well as trace amounts of other essential elements. The fertility of the Nile delta was replenished as these fine particles settled on the flooded fields.

This annual gift of humus from the forests of the upper Nile was taken for granted until high dams built on the upper Nile held back the floodwaters and trapped the humus before it could be carried off to the delta. Nowadays what little farming is done in the Nile delta must be done with the expensive and ephemeral help of fertilizers. The other great river valleys where early civilizations began with such promise eventually met a similar fate. Once the forests along the rivers had been cut and the spongy soil on which they had grown began to wash away, little humus and organic matter remained to restore fertility to the fields in the valleys.

Plato's dialogue, *The Critias,* dating from the fourth century BC, describes the sad state of what was once rich soil in terms that have become all too familiar in the twenty-first century.

> What now remains of the formerly rich land is like the skeleton of a sick man, with all the fat and soft earth having wasted away and only the bare framework remaining. Formerly, many of the mountains were arable. The plains that were full of rich soil are now marshes. Hills that were once covered with forests and that produced abundant pasture now produce only food for bees. Once the land was enriched by yearly rains, which were not lost, as they are now, by flowing from the bare land into the sea. The soil was deep, it absorbed and kept the water in the loamy soil, and the water that soaked into the hills fed springs and running streams everywhere.

6. Discouraging Invasion of Soils by Exotic Species

Disturbed areas with soils that have lost their structure from erosion, compaction, and clear-cutting are most susceptible to rapid invasion and establishment of exotic plant and animal species. Invasive species can dramatically disrupt the normal interactions and species composition of the underground food web and the aboveground food web that are so intimately linked.

Exotic flatworms introduced to Europe and North America are reducing earthworm populations in their adopted habitats, while exotic earthworms are depleting leaf litter of some North American forests at rates far exceeding the recycling of leaf litter carried out by native earthworms. As mentioned in the entry on earthworms (p. 77), the rapid disappearance of this leaf litter from forest floors is endangering the lives of many plants and animals that dwell in this leaf litter.

Many introduced species of plants are invading different soil habitats around the world, displacing local species, and disturbing the associations of once stable soil communities. Garlic mustard is one of many invasive species introduced to North America forests from Europe and one of the very few plants that does not establish mycorrhizal associations. Chemicals released into the soil by this plant are responsible for disrupting the mycorrhizal associations of fungi and trees, with far-ranging, but still poorly understood, implications for the diversity and well-being of other life underground and aboveground.

7. Composting as an Antidote to Soil Abuse

Rotten to the core. This quality just comes naturally to a compost pile and endears it to the decomposers and the company they keep. You can create a refuge for these creatures in your backyard, help solve the world's garbage problem, and provide a bountiful supply of organic nutrients for your garden by becoming a composter. All plant matter contains nutrients from the soil in which it grew. To dispose of any plant materials in a landfill, whether they are lawn clippings, fallen leaves, produce from the grocery, or even weeds, is to waste a wonderful opportunity to give back to the soil some of the nutrients and organic matter that would otherwise be squandered (plate 66).

By participating in nature's processes, not only will you feel good about your contribution, but the fruits, vegetables, and flowers of your garden also will benefit from the improved structure that compost gives to the garden's soil as well as the extra nutrients and moisture that will come their way. Your compost will be a refuge for many small and fascinating creatures that will quickly and unobtrusively move in to carry out their job of recycling, renewal, and restoration.

By composting we create a hospitable environment for many of the soil's decomposers so they can move in and carry out the same tasks that they carry out among the plant debris on the forest floor, in a meadow, and even in all but the most heavily sprayed and fertilized lawns and fields. Once the decomposers have arrived, the conditions in the compost pile encourage them to stick around and multiply.

Composting speeds up the natural process of decomposition that occurs constantly wherever weeds die, leaves fall, and logs rot. Composting is based on four simple and interdependent principles. By (1) interspersing green, moist matter with brown, dry matter and by (2) constructing a pile that is just the right size, a composter (3) assures that air circulates through the compost and (4) that the compost maintains a moisture content that best suits the needs of the decomposers. But even if it is not possible to adhere to all of these principles, your compost will still decompose, although it might take longer to do so. Nature always seems to compensate for imperfections and will complete the job of decomposition at its own pace.

With only the first and the last lines of his poem *This Compost,* Walt Whitman conveyed this remarkable transformation of compost that is orchestrated by decomposers. "Behold this compost! Behold it well! . . . It gives such divine materials to men, and accepts such leavings from them at last."

The few steps outlined here will give you a good idea how simple, straightforward, and rewarding composting can be.

(a) Begin by finding a good location for the compost pile. Choose a place away from walls and fences that can decay and stain. Avoid any spots that have poor drainage; but if you can locate the compost near a source of water and within reach of a hose, you will have the advantage of easily maintaining the optimal amount of moisture for composting.

(b) Learn what sort of items the decomposers of the compost pile find acceptable. As long as the items are organic and biodegradable or inorganic and nourishing, the decomposers can extract sufficient nutrition and energy for their growth. However, avoid adding meat scraps that might attract rodents and pets to the compost pile. Also leave out rags and tires, charcoal and coal ashes, diseased plants, sawdust from treated lumber, and pet litter; even though they are organic, they either do not degrade well or can contain toxins as well as harmful organisms. Shred or pulverize large and hard materials like corn cobs, stalks, shells, and fruit rinds before adding them to the pile. Chopping up these items will accelerate their decay by increasing the surface area on which bacteria and fungi can work.

If we made a detailed list of items, we would soon see that the items could be grouped into general categories: (1) plant matter—either green, dried, or processed (including shredded newspapers, coffee grounds, wood ashes, and wood chips); (2) mineral matter such as ground stone or phosphate rock; (3) dried animal matter such as feathers, bone meal, blood meal, fish meal, hoof and horn meal, eggshells, and shellfish; and, of course, (4) a variety of manures.

Each new addition to the pile contains a different mix of nutrients. Some, like seaweed, are rich in the essential trace element boron; onion and cabbage scraps are rich in sulfur; and shells of all sorts are excellent sources of calcium. The greater the variety of items placed on a compost pile, the greater the diversity of microhabitats you will provide for decomposers, and the richer and more balanced the mix of nutrients in your compost.

What is important to remember is that the decomposers require nutrients of their own to provide the energy for their survival and work. Microbes need a certain proportion of carbon to nitrogen in order to multiply and carry out the job of decomposing the compost; usually a carbon:nitrogen (C:N) ratio somewhere between 15:1 and 30:1 is ideal.

Nitrogen is the nutrient that is almost invariably in short supply. Nitrogen added to compost in the form of manure, fertilizer, or green vegetation such as weeds and grass clippings provides the missing element for microbial growth and really speeds up the decay of the compost. Fresh, green plant matter is higher in nitrogen than dry, dead plant matter like straw, sawdust, or dry leaves and is always a good addition to the compost pile. The following C:N ratios for various materials can serve as a useful guide in choosing the best nutrition for the decomposers in your compost pile.

Garden soil	12:1–15:1
Manure	15:1–20:1
Grass clippings or weeds	20:1–25:1
Dry leaves or straw	30:1–60:1
Pine needles	60:1–100:1
Sawdust or wood chips	150:1–700:1

(c) Collect a critical mass of organic materials. Use leaves, grass, weeds, or kitchen scraps for optimal composting. Decomposers thrive best on about equal portions of carbon-rich and nitrogen-rich materials. These should be added to the pile in alternating layers, each about four inches (10 centimeters) thick. Experienced composters often have a separate bin for stockpiling materials before adding them to an actively decomposing pile. Compost will heat up more quickly and break down more rapidly if materials are collected in advance and then added simultaneously.

(d) Decide on what form the compost pile should take. You can simply heap compost on the ground and let gravity and the decomposers mold the shape of the pile. Books on composting, however, describe pits, bins, barrels, tumblers, boxes, steel drums, garbage cans with holes in their bottoms and sides, and even plastic bags for storing and processing compost. If space is limited, begin your composting in a trash can, large plastic bin, or a large wooden box. Add several holes to the container to make sure that air can circulate around and through the compost.

(e) Decide how large the pile should be. A pile that is too small will lose a large amount of the heat generated by the microbes as they rapidly break down or oxidize carbon compounds to carbon dioxide. However, a pile that is too large may become too hot or too airtight at its core. High temperatures generated inside many compost piles can reach 165° F (74° C) as bacteria, actinomycetes, and fungi reach population densities of 10 billion microbes for every gram of compost. The heat may kill off some of the decomposers, and the lack of oxygen will eliminate many of the aerobic microbes that are most efficient at decomposition. High temperatures speed up the process and destroy plant pathogens as well as the seeds of any weedy plants that happened to find their way into the pile (plate 67).

(f) Add a "starter" or an "activator" to speed up the arrival of decomposers in your compost. Starters are always rich in nitrogen compounds that promote the rapid growth of bacteria and fungi. In addition, many starters such as manure, well-decomposed compost, or rich soil also come with their own populations of bacteria, fungi, and some larger decomposers. Bread makers use the same strategy when they add a starter of yeast to their bread dough. In the same way that different yeasts contribute different flavors to the bread, different compost starters add different amounts of nitrogen-rich matter as well as different populations of decomposers.

(g) Make sure the microbes in your compost have plenty of air. A well-aerated compost undergoes thorough decomposition to carbon dioxide, water, minerals, and humus as opposed to the partial, smelly decomposition that occurs in compost that is unmixed and airtight. A layer of coarse material such as cornstalks or tomato stalks at the

very bottom of a pile allows air to enter the pile from below. Turning the pile about a week after it has reached its highest temperature will improve the air supply and hasten the decomposition.

Contrary to popular belief, a compost heap that is properly tended does not stink. As long as the bacterial and fungal decomposers in the compost pile are getting enough oxygen, they continue adding oxygen to carbon-containing compounds until most of the carbon in these compounds has been converted to carbon dioxide. This conversion process is known as oxidation, and the microbes that carry out the process are known as oxidizers or aerobes. They have a real talent for quickly breaking down dead plant material and dead animal matter into carbon dioxide, minerals, humus, and water.

During this chemical transformation, considerable energy is given off in the form of heat. The very same basic components—carbon dioxide, water, minerals, and energy—that green plants start out with during photosynthesis as they form the organic compounds of leaves, wood, and fruit eventually are generated in a compost pile. The pathway followed by these components comes full circle, as they pass from living plants and animals to the decomposers of the soil and then back to plants and animals again.

However, if the breakdown of compost is turned over to the bacteria and fungi that can survive in the absence of oxygen, these anaerobes (*an* = without; *aero* = air) ferment large organic compounds to smaller organic compounds without completely converting them to carbon dioxide.

During bread making and wine making, anaerobic microbes are the ones that convert sugars to alcohol along with some carbon dioxide. These anaerobic microbes also decompose plant and animal debris to compounds of carbon like butyric acid, lactic acid, or alcohol as well as stinky gases such as hydrogen sulfide and ammonia. Compost piles should clearly be managed to encourage aerobic decomposers rather than their anaerobic relatives.

(h) Keep the compost moist, but not wet. Water is another essential requirement for composting. As mentioned earlier, locating the compost pile near a hose or water source is a good idea. The decomposers do their best work when the compost is uniformly moist but not waterlogged. Accomplished composters say that the pile should have the water content of a squeezed sponge.

Nature keeps the generation of humus simple and there is no reason why composting should not be a simple process as well. Nature carries out the processing of organic matter at its own pace and sees to it that the populations of various bacteria, fungi, actinomycetes, and larger decomposers coexist in the proper proportions. You can sit back and watch as the decomposers work their magic in the backyard compost or in the litter underfoot.

In a journal entry from 1856, Thoreau summarized Nature's approach to recycling. "In Nature nothing is wasted. Every decayed leaf and twig and fibre is only the better suited to serve in some other department, and all at last are gathered in her compost heap."

Healthy soil, healthy food, and healthy people are inextricably linked. Wherever good stewardship of the land is practiced—by controlling erosion and acid rain, by minimizing use of pesticides and tillage, as well as by using compost and manures instead of commercial fertilizers—we benefit by having richer soils and more nutritious food. At the same time, these practices preserve habitat for soil creatures, our partners in assuring that the gift of good earth will be cherished and not squandered.

Collecting and Observing Life of the Soil

If at all possible, observe living creatures of the soil as they go about their business. Some of the observation techniques described in this section can be used in the field, but at times close observation requires study indoors. Rachel Carson would take creatures from the seashore to observe in her study, but she would always return them to the places where she had collected them. Try to observe this same courtesy with the creatures that inhabit the soil.

Every creature is better alive than dead, men and moose and pine trees, and he who understands it aright will rather preserve its life than destroy it.

HENRY DAVID THOREAU

WHAT TO USE FOR OBSERVING

Without using any devices for magnification, the human eye can spot many details in the world around it; but discovering unseen details of that world with each increase in magnification offers the experience of exploring new worlds within worlds. Magnifiers of many different sizes and different powers of magnification are available, but my favorite magnifier for viewing the creatures I find in the field is a stereomicroscope with zoom magnification that allows the viewer to quickly switch, or zoom, from low (~10×) to high (~50×) magnification. Individual creatures or communities of creatures in such natural environments as a decaying leaf or a clod of soil are viewed using illumination either from below or from above the stage on which the

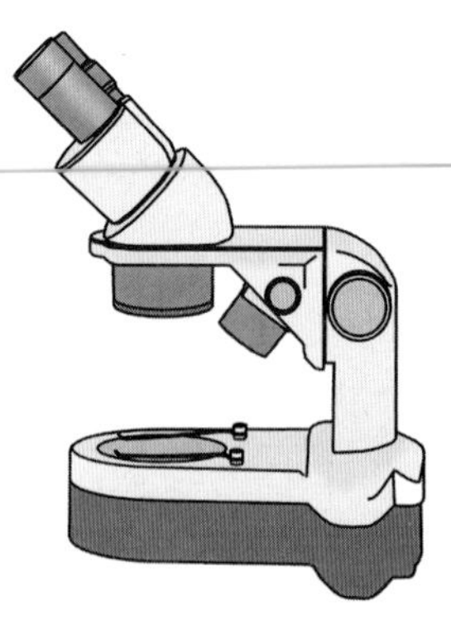

208. Examining soil creatures through a stereomicroscope offers rarely seen views of a dark and hidden world.

creatures are placed. These stereomicroscopes are reasonably priced; many classrooms and nature centers now have them (fig. 208).

WHERE TO LOOK IN THE FIELD

Finding Microbes

Microscopic examination of decaying leaves and logs reveals a miniature forest of fungal hyphal threads and fruiting bodies that can be simple, branched, saucer-shaped, or pot-shaped. The shells of testate protozoa are often evident. Sometimes the bright yellow plasmodium of a slime mold shows up.

Fungi

Only an inch or two beneath the soil surface in a beech or oak forest lie the distinctive short and stubby roots of beech trees or oak trees that are covered by ectomycorrhizal fungi. Although these roots are very abundant and easy to find, very few people have ever seen the mycorrhizal marriage of roots and fungi (fig. 209).

Gardeners can bait for soil fungi such as *Metarhizium* and *Beauveria* that infect insect pests. Place a dead insect in a moist chamber with a sprinkling of garden soil. Wait a few days to see if the insect is covered with mold that can grow on hard insect cuticles. Fungi that are pathogens of insect pests will colonize a healthy soil that contains an abundant supply of humus.

Bacteria

Individual soil bacteria are exceedingly small and visible only when viewed with the high magnification of a compound microscope or an

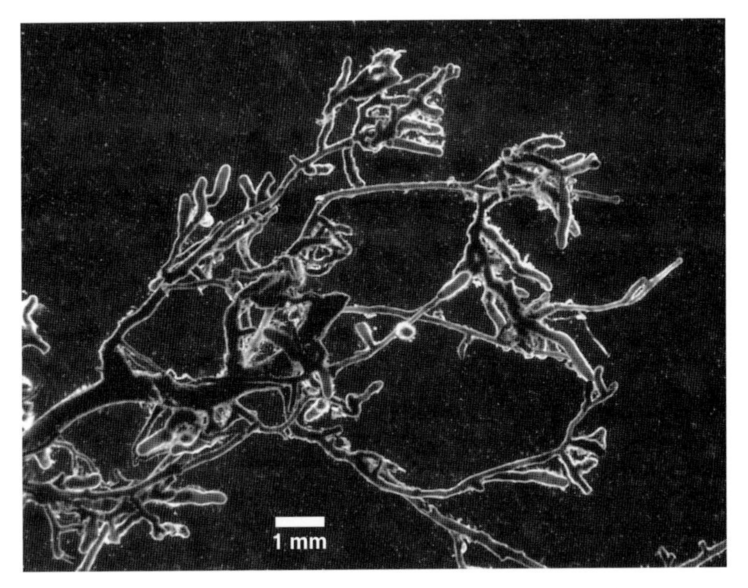

209. The distinctive short, club-shaped root tips of oak mycorrhizae lie just below the surface of the forest soil.

electron microscope. In order to convey the diminutiveness of these microbes, Sir John Russell pointed out in his book *The World of the Soil* (1957) that most bacteria are so small that "a quarter of a million of the organisms could sit comfortably on an area the size of the full stop at the end of this sentence."

But collectively these millions of bacteria often leave obvious traces of their whereabouts in the field. These bacterial field marks, as they are referred to in *A Field Guide to Bacteria* by Betsey Dexter Dyer, can be easy to spot. Pink salt crusts indicate the presence of salt-loving archaebacteria; and rusty, red soils are associated with iron-oxidizing bacteria. The rotten-egg odor of dark, anaerobic soils can be traced to sulfur-reducing bacteria; and bubbles of methane produced by methanogenic bacteria rise from anaerobic soils underlying stagnant pools of water.

Legumes are some of the most common plants of roadsides and impoverished soils. The roots of clovers and other legumes such as peas and beans are often densely covered with nodules filled with millions of rhizobial bacteria. The bacterial enzyme that fixes nitrogen must carry out its reaction in the absence of oxygen. To assist the enzyme, plant cells of these nodules produce a red protein similar to human hemoglobin that binds oxygen and prevents it from interfering with the function of the nitrogen-fixing enzyme. Look for these colored nodules on small, young roots of legumes (see plate 4).

Finding Invertebrates

Turn over rocks, bricks, boards, logs, or other objects that have lain on the ground for some time to see a two-dimensional view of life underground. Against the dark background of the underlying soil, you can often readily spot lightly pigmented animals, even very tiny ones. Notice how they move. The springtails will hop about. Earthworms will retract into their burrows. The centipedes will speed off to a dark recess, and the woodlice will lumber off or curl up in a ball. Always carefully return the rock or log to its original location so that the lives of creatures under these logs and rocks can return to their normal routine after this sudden intrusion.

Mother skunks teach their kits to probe through cow pads for the insect meals that lie within. Probing around in cow pads of different ages can be a rewarding lesson in ecological succession. The initial colonizers of a freshly deposited cow pad are succeeded by a steady stream of different creatures—worms, mites, fungi, insects—until the last crumbs of the pad disappear in the soil. Which creatures are scavengers or fungivores and which are predators?

The animals that live in a soil or leaf litter sample are usually so abundant that they can be found just about anywhere by closely and patiently observing. However, a few simple methods have been developed to extract animals from soil and leaf litter that can be used to greatly increase the number of species and the number of individuals that you are able to find wherever you happen to look beneath your feet.

There are a number of ways to extract soil invertebrates from their habitats.

Rearing creatures from soil and litter samples. One way to observe ecological succession in decaying cow pads, rotting mushrooms, decaying leaf litter, or decaying wood is to place the sample in a tightly sealed cardboard box whose bottom has been lined with a plastic sheet to prevent moisture of the decaying sample from promoting the rotting of the cardboard as well. But still add water periodically to ensure

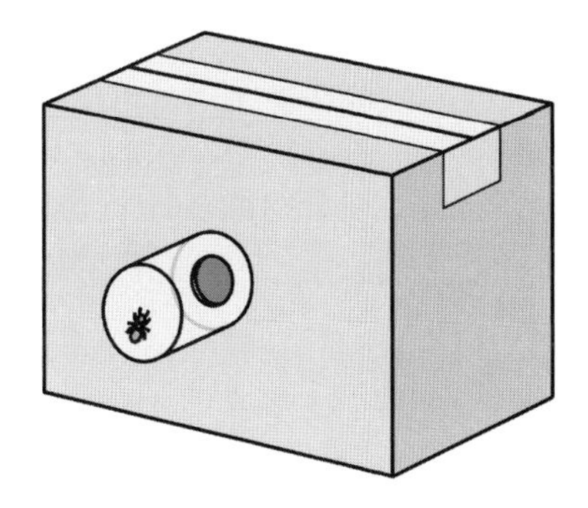

210. From samples of soil, leaf litter, or decaying logs that have been placed in dark, damp emergence boxes, adult insects will emerge and head for the closest and only source of light.

that the sample does not dry out. At one end of the box, cut a round hole in which the mouth of a small jar is barely but snugly inserted. As adult insects emerge from the samples, they will head for the only source of light in the box and congregate in the jar (fig. 210).

Sifting through soil and leaf litter. A large kitchen strainer or a sieve made from hardware cloth screen can be used in the field to sift creatures from soil and/or leaf litter. Sifting through a soil sample will concentrate creatures above a certain size in the strainer as soil particles pass through. Creatures below a certain size in a litter sample will pass through the strainer as litter is retained. Bring along a white sheet or a white enamel tray, where the creatures can be examined and sorted (fig. 211).

Flotation. A white pan or tray, a beaker or a bottle, and a bucket of water are all the supplies you will need to try this simple field method for separating arthropods from soil. Because most arthropods are less dense than water, they will float to the surface after the soil in which they are living has been suspended in a bucket or large beaker of water. Skim off the surface water with the floating arthropods into a small beaker. Place the skimmed water in a white pan or on white filter paper in a dish or pan. Most of the arthropods will be mites, springtails, proturans, pseudoscorpions, and other tiny invertebrates; they will be intermixed with small particles of organic matter from the soil. Watch for movement, and as your eyes zoom in on particular creatures, you'll soon be able to distinguish the different groups that are represented in the soil sample (fig. 212).

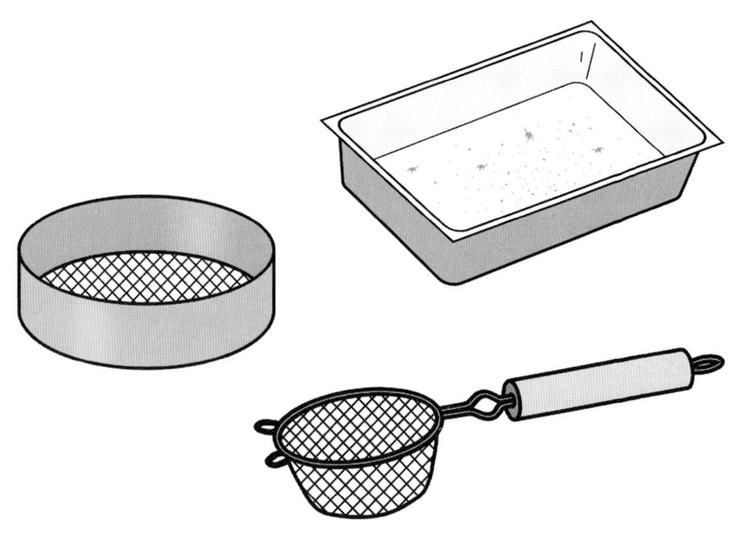

211. A large kitchen strainer or sieve made from hardware cloth screen will separate large volumes of soil or leaf litter from some of the animals that live in those habitats. Even tiny animals that have been sifted from their habitat can be conspicuous when placed on a white sheet or in an enamel pan.

212. When a soil sample is thoroughly mixed with water, most soil arthropods in the sample will float and can be skimmed off the water's surface.

Using an aspirator. Small arthropods are difficult to handle with fingers or even forceps. The best way to move these small animals from place to place is to use a simple device made with a two-hole stopper, some glass tubing, a length of flexible rubber or plastic tubing, and a glass or plastic jar. Approach a small arthropod with the tubing protruding from the jar while you are inhaling air from the tub-

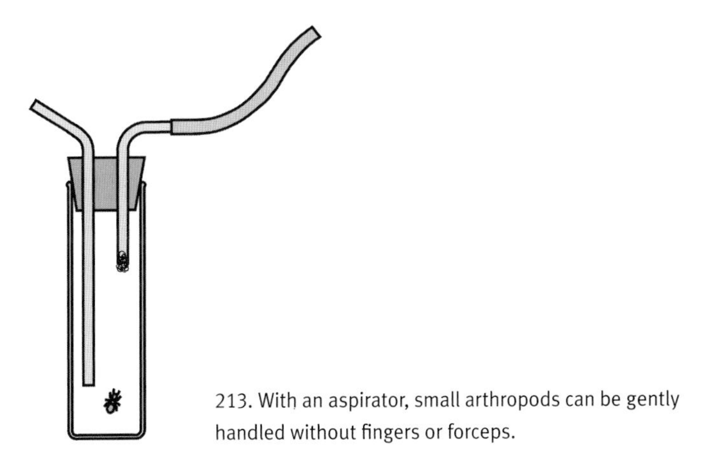

213. With an aspirator, small arthropods can be gently handled without fingers or forceps.

ing inserted in the jar. The act of inhaling draws air into the jar along with the arthropod. Several arthropods can be gathered in quick succession with an aspirator (fig. 213).

Berlese funnel. Another way to make the acquaintance of those that live beneath our feet is to set up a Berlese funnel. This useful device is named after the Italian scientist Antonio Berlese, from whose writings and beautiful, detailed drawings came some of our first close-up views of the subterrranean world. The heat and brightness of an electric light placed above a Berlese funnel drives creatures that are accustomed to dark, moist habitats down into the dark chamber below the funnel. Here they can be collected and observed at close range. This method of extracting soil arthropods from their habitat is probably the most thorough of all the methods. It is hard to believe that a device so simple to make can be so effective at revealing who lives in the leaf litter and soil (figs. 214–15).

Baermann funnel. The variety of nematodes, potworms, protozoa, and rotifers in a handful of soil can be easily sorted from humus and mineral particles by wrapping soil in cheesecloth before placing it in a small funnel that is filled with water after its stem has been corked or clamped at its end. In a short time most of the invertebrates wriggle out of the inundated soil, through the pores of the cloth, and into the water. They soon gravitate to the stem of the funnel; and when the

Berlese Funnel

Large funnel
Leaf litter or soil sample
Top of plastic soft drink container
Mesh or screen to hold sample
Mesh
Dark container
Bottom of plastic soft drink container covered with foil to keep the interior dark
Alcohol or water in collecting dish

214. Berlese funnels are easy to make and use. Sooner or later all the arthropods of a soil or leaf litter sample leave the funnel to escape the bright light overhead and look for a darker and more humid home.

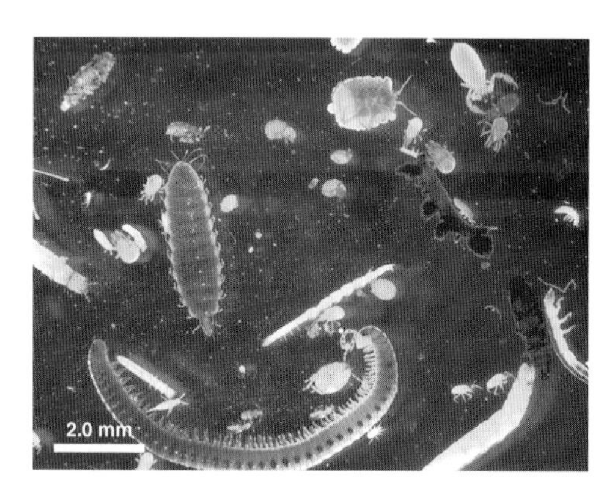

215. A sampling of the arthropods that live in a handful of leaf litter. These insects, mites, springtails, a millipede, and a pseudoscorpion were recovered with a Berlese funnel.

cork or clamp is removed, many creatures can be collected for closer inspection under a hand lens or microscope (figs. 216–17).

Pitfall traps. These traps collect live animals that wander about on the soil surface. The traps can take the form of bottles, cans, or plastic cups that are placed in the soil with their openings at ground level. Place a slightly elevated flat rock or board as a cover for the opening. The space between the cover and the opening of the pitfall should be wide enough for large insects to enter but small enough to keep out small mammals. The cover often can help protect the trap from marauding animals such as raccoons or from flooding during a rain. Be sure to check the traps at least once each day and always remove the traps when you are finished. Frequent inspection of the traps

Baermann Funnel

Cheesecloth or muslin bag with soil sample
Funnel
Top of plastic soft drink container
Water
Rubber tubing
Clip
Cap
Ring stand
Bottom of plastic soft drink container
Collecting dish

216. Baermann funnels extract protozoa, nematodes, potworms, and other invertebrates from a handful of soil.

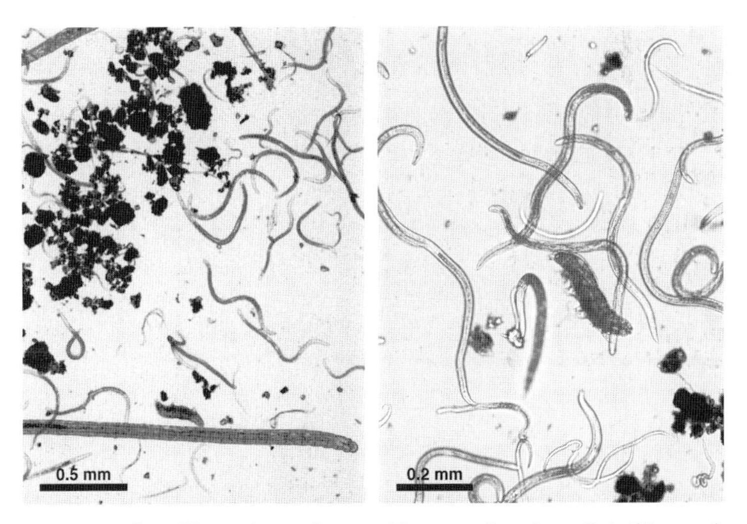

217. A sampling of invertebrates from a tablespoon of garden soil. In this sample from a Baermann funnel, a tardigrade and a potworm mingle with many nematodes.

will ensure that trapped invertebrates can be eventually released unharmed (fig. 218).

Pitfall traps can be used to estimate the abundance of certain arthropods in a given area. A clever and simple method was developed that can provide this information. Set up a series of pitfall traps in a field or forest that you will check the next day. Mark the individual

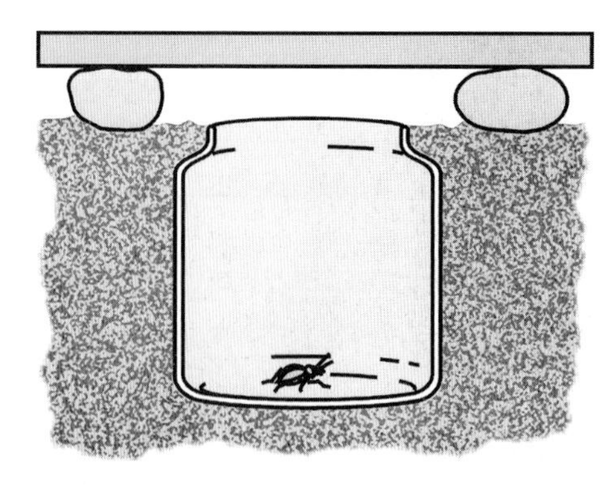

218. Pitfall traps capture creatures that roam about on the surface of the soil.

animals whose population you wish to estimate. Apply a tiny spot of quick-drying, waterproof paint to each of the animals and then release them. When you return the next day to check the traps, note how many of the marked individuals were recaptured. A good estimate of the population size of the animal you are studying is

$$\text{approximate size of population} = M_1 \times M_2 / N_2$$

where

M_1 = number of animals captured and marked on day 1
M_2 = number of animals captured on day 2
N_2 = number of animals marked on day 1 and recaptured on day 2.

OBSERVATION CHAMBERS

In his book *The Soil and Health* (1947), Sir Albert Howard describes how a glass window that had been installed along one side of a pit in the soil of an orchard provided a window on the world of roots and the mycorrhizal fungi that surround them like so many "cobwebs." Six-sided observation chambers can be built for study indoors. These chambers work well for many soil inhabitants—plant roots, animals, fungi. People most frequently use them for ants, or termites, or earthworms; but these chambers, when filled with soil and leaf litter, can be homes for just about any soil creatures. By keeping these chambers covered or stored in a dark room when they are not being observed, the creatures of this dark and hidden world should remain healthy and content (fig. 219).

Small fungus-eating arthropods, such as mites, springtails, and proturans, as well as predatory arthropods, such as pseudoscorpi-

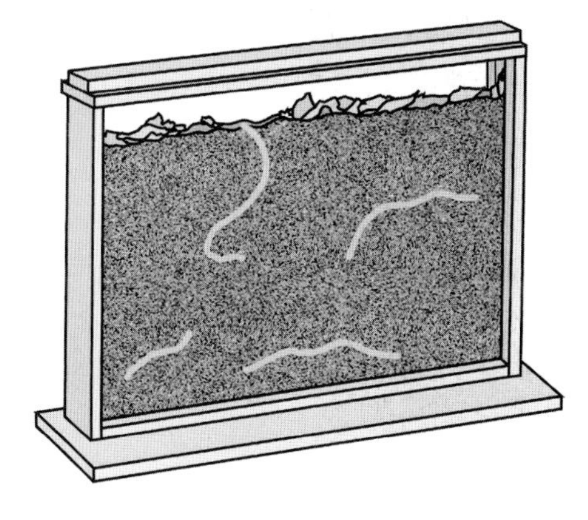

219. Observation chambers can be built to hold a segment of underground habitat and the creatures that are found there.

220. Smaller observation chambers for mites, springtails, and some of their predators simulate the moist environment of soil. A porous plaster of Paris substrate in a small dish or jar holds moisture required for the survival of the arthropods and the yeast on which most of them feed.

ons, diplurans, and beetles, can be reared at home or in a classroom. Pouring a layer of plaster of Paris into a small glass or plastic jar or dish will provide a porous surface that can retain moisture for long periods of time. Adding powdered charcoal to the plaster often enhances the contrast between the arthropods and their substrate and helps in observing those that have little or no pigment. The fungivores will grow and multiply on a few pellets of dry baker's yeast that are added periodically. Add water frequently to the porous plaster and cover the container to retain moisture (fig. 220).

ATTRACTING TOADS, LIZARDS, SNAKES, AND BIRDS TO A GARDEN

A garden rich in mulch and compost will provide an enticing habitat for many invertebrates as well as the vertebrates that feed upon them. These vertebrates will find the garden even more appealing if the area includes a pool, a bath, or a fountain. The insects, earthworms, and mollusks that reside in the garden can attract many spe-

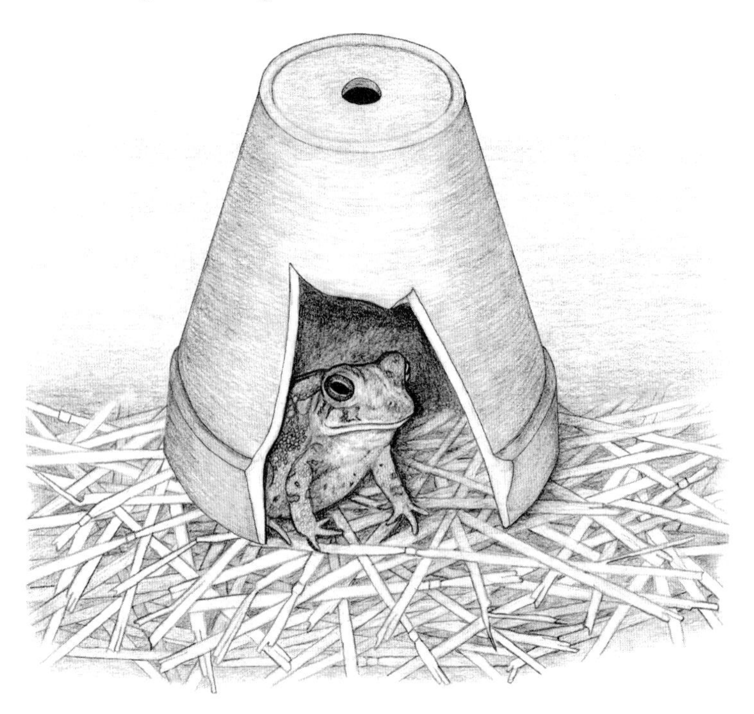

221. Gardens can be made more attractive habitats for toads by providing them with old clay pots for shade and shelter.

cies of birds, a number of small, harmless snakes, toads, and lizards. A greater diversity of vertebrate species will be found in gardens at warmer latitudes.

Each nest box or nesting platform placed in a garden provides a home for an entire family of birds. Rocks that are part of a garden landscape can provide cover for snakes and lizards as well as places for basking in the sun. Toads soon take up residence in the shelter of old, broken clay pots placed in shaded areas of the garden (fig. 221).

Glossary

Acid soils: Soils having a concentration of hydrogen ions greater than one part in 1 million. *See also* alkaline soils, neutral soils.

Actinomycetes: Microscopic organisms that have attributes of both bacteria and fungi. They form branched, multicellular filaments and look like fungi, but their genes resemble bacterial genes. The rich odor of moist, freshly plowed earth emanates from the hordes of actinomycetes that inhabit the soil.

Adsorption: Attraction of compounds or ions to the surface of a particle.

Aerobes: Organisms that survive and grow only (obligate) or usually (facultative) in the presence of air or oxygen gas.

Aggregate: Many soil particles that are held together in a single mass, such as a granule, a crumb, or a clod.

Algae: Autotrophic organisms that use the energy of sunlight captured by chlorophyll to produce sugars from carbon dioxide and water. These simple soil organisms without stems, roots, or leaves are considered members of three different kingdoms: the Plants (green and red algae), the Eubacteria (blue-green algae), and the Chromista (algae with flagella, yellow-green algae, golden algae, and diatoms).

Alkaline soils: Soils having a concentration of hydrogen ions less than one part in 100 million. *See also* acid soils, neutral soils.

Anaerobes: Organisms that survive and grow only (obligate) or usually (facultative) in the absence of air or oxygen gas.

Anions: Negatively charged elements or compounds.

Arachnids: Arthropods that have four pairs of legs, no antennae or wings, and only two main body regions (cephalothorax and abdomen) instead of three regions (head, thorax, and abdomen). Among the arachnids are spiders, mites, daddy longlegs, and pseudoscorpions.

Arthropods: A large group (phylum) of animals with jointed appendages and no backbones.

Autotrophs: Creatures that obtain the energy that they need for survival from either the sun or from the reactions of inorganic (mineral) components of the soil. The carbon that these creatures need to form organic compounds comes from carbon dioxide.

Biomantle: The uppermost portion of the earth's mineral soil, inhabited and molded by living creatures.

Casts: Nutrient-rich aggregates of soil that are the droppings of earthworms.

Caterpillar: The larva of a moth or butterfly.

Cations: Positively charged elements or compounds.

Cellulose: Long chains of sugar molecules found in all plant tissues. The sugar molecules are always glucose units. Cellulose forms fibers that support and strengthen these tissues.

Cerci: A pair of sensory appendages on the tenth segment of an arthropod's abdomen that are sometimes used as pincers.

Chelicerae: The jaws of arachnids that can take the form of fangs, pincers, or beaks.

Clay: A class of inorganic soil particles that measure less than 0.002 mm in diameter, formed by the chemical weathering of rocks such as granite.

Composting: The act of generating humus from organic materials that have been collected and mixed outside of the soil, where they decompose with minimal loss of nutrients.

Compounds: Chemicals that are made up of two or more elements united in constant proportions.

Coniferous: Pertaining to trees and shrubs that produce cones and are usually covered with evergreen leaves shaped like needles or scales.

Coprophages: Organisms that feed on dung and droppings of other creatures.

Cryptobiotic soil: An apparently lifeless soil that is actually teeming with microbial life—bacteria, protozoa, algae, lichens, fungi—and sometimes simple plants such as mosses. The occupants are the first colonizers of this apparently barren landscape.

Deciduous: Pertaining to plants that shed their leaves each year at a certain season.

Decomposer: An organism that obtains nutrients by breaking down the remains or waste products of other organisms.

Denitrification: The release of oxygen from nitrates (NO_3^-) or nitrites (NO_2^-), resulting in the formation of nitrogen oxide gases (NO or N_2O) or dinitrogen gas (N_2); ultimately these nitrogen gases are lost from the soil.

Detritivore: An organism that feeds on dead plant and animal matter (detritus).

Ectomycorrhizae: *See* Mycorrhizae.

Elements: Chemicals that cannot be broken down to other chemicals with different properties. Some elements used by plants are obtained from air and water, but most elements are derived from the minerals of soil.

Endomycorrhizae: *See* Mycorrhizae.

Euedaphic: Pertaining to the inhabitants of the dark, deep layers of the soil.

Eukaryotic: Pertaining to cells with a true nucleus. Such cells are found in all organisms except the bacteria in the kingdoms Eubacteria and Archaebacteria.

Extremophiles: Organisms that inhabit extreme environments that can be hyperthermal, hypersaline, or anaerobic. The term is often applied to members of the Archaea.

Fossorial: Adapted for digging.

Fungi: Organisms with spores and filaments (called hyphae), once considered plants and now placed in their own kingdom; they can exist either as single cells or multicellular organisms. While some feed on other living creatures, most are decomposers.

Fungivore: An organism that feeds on fungi.

Gastropod: A snail or slug.

Greenhouse gas: A gas such as carbon dioxide, methane, or nitrous oxide that traps heat in the upper atmosphere of the earth just as glass traps heat in a greenhouse.

Hemicellulose: Long chains of sugar molecules found in all plant tissues. The sugar molecules that make up hemicellulose chains consist not only of glucose units but also of other sugar units such as mannose, xylose, and galactose. Along with cellulose, hemicellulose forms fibers that support and strengthen plant tissues.

Herbivore: An organism that feeds on plants.

Heterotrophs: Creatures that obtain their energy and carbon from the decomposition of organic components produced directly or indirectly by autotrophs.

Horizon: A layer of soil.

Humus: The dark organic matter that remains after most plant and animal debris has decomposed.

Hyphae: Filaments or strands of fungal cells.

Immobilization: The transformation of an element (nutrient) from its inorganic form in the soil to organic forms within a plant or microbe. Once incorporated within living tissue, the element is no longer readily available to other plants or microbes of the soil.

Inorganic matter: Any chemical or material that does not contain compounds having both carbon and hydrogen.

Invertebrates (*in* = without; *vertebra* = segment): Multicellular animals without backbones.

Larva: An immature insect or other invertebrate that shows little resemblance to its adult form. If the adult insect has wings, these wings first form inside the body of the larva.

Leaching: Removal of nutrients from a soil layer by water's passing through that layer.

Legume: A member of one of the three largest families of flowering plants that include peas, beans, and clovers. Rhizobial bacteria form nodules on the roots of legumes, where they convert dinitrogen gas from the air to ammonia.

Lichen: A composite organism that is part alga and part fungus. The alga and fungus live together in a cooperative arrangement, often surviving in habitats where neither partner could survive alone.

Lignins: Large organic molecules that impart rigidity to plant tissues and are particularly resistant to the decomposition that accompanies formation of humus. Lignin molecules hold together cellulose and hemicellulose chains.

Loam: A soil texture to which sand, silt, and clay contribute almost equally.

Maggot: The larva of flies in the suborder Cyclorrhapha of the order Diptera.

Manure: Animal droppings containing organic matter as well as inorganic nutrients that enrich the soil by their addition.

Microbes: Organisms that cannot be easily viewed without a microscope. Among them are bacteria, actinomycetes, fungi, algae, and protozoa.

Mineralization: The conversion of elements in the remains of organisms to their inorganic forms.

Minerals: Inorganic compounds derived from rocks or from the remains of organisms. Some rocks consist of a single mineral (such as limestone, which consists solely of calcium carbonate); other rocks such as granite represent mixtures of minerals.

Mor: A type of humus layer that is clearly separate from the underlying mineral layers of soil. Mor is typical of coniferous forests.

Mulch: A layer of organic matter of plant and/or animal origin that blankets the soil, protecting it from erosion, retaining moisture, and adding nutrients.

Mull: A type of humus layer that mixes with the underlying mineral layers of soil. Mull is characteristic of deciduous forests.

Mycorrhizae: The mutually beneficial association of fungi with roots of plants. Fungi of **endomycorrhizae**, or **vesicular-arbuscular mycorrhizae**, have hyphae that actually penetrate the walls of root cells but not the membranes of these cells. Fungi of **ectomycorrhizae** have hyphae that surround each small tree root and radiate into the surrounding soil. Hyphae of ectomycorrhizae also penetrate roots by passing between cell walls, establishing a network of filaments among the root cells without ever penetrating their cell walls or cell membranes.

Neutral soils: Soils having a concentration of hydrogen ions around one part in 10 million. *See also* acid soils, alkaline soils.

Nitrification: The conversion of ammonium (NH_4^+) first to nitrite (NO_2^-) and then to nitrate (NO_3^-) by the addition of oxygen. The first step is carried out by autotrophic bacteria called *Nitrosomonas,* while the second is carried out by other autotrophic bacteria called *Nitrobacter.*

Nitrogen fixation: The conversion of dinitrogen gas of the atmosphere to nitrogen compounds that can be used by living organisms.

Nutrients: Elements or compounds that nourish and promote the growth of organisms.

Nymph: An immature insect that usually resembles its adult form. If the adult has wings, these wings first form on the outer surface of the nymph's body.

Organic matter: Plant or animal material found in various stages of decay. Organic matter always contains compounds of the elements carbon and hydrogen.

Oxidation: The loss of negative charge by a substance, often accompanied by chemical combination of the substance with oxygen.

Parent materials: The rocks and organic matter from which soil originally forms.

Pedipalps: A pair of sensory appendages on the head of an arachnid; the counterpart of an insect's antennae, sometimes used as pincers.

Photosynthesis (*photo* = light; *syn* = together; *thesis* = an arranging): The process by which plants, algae, and certain bacteria use light, carbon dioxide, and water to produce sugars and oxygen.

Podzol: A type of soil that forms under coniferous forests and has a characteristic bleached horizon from which nutrients have been leached by precipitation, leaving a pale layer of sand grains.

Producers: Plants, algae, and autotrophic bacteria that produce organic nutrients from simple inorganic compounds using energy from sunlight or from the reactions of inorganic compounds.

Prokaryotic: Pertaining to cells and microbes lacking a true nucleus.

Protists: Eukaryotic microbes that include protozoa and members of the kingdom Chromista.

Protozoa: Single-celled organisms that include amoebae with and without shells, those having flagella (*flagellum* = whip), and those having cilia (*cilium* = small hair).

Pupa: The stage in the life cycle of some insects that lies between the larval and adult stage.

Puparium: The hard exoskeleton of the last larval stage of certain flies that is not shed but covers and protects the pupa.

Reduction: The gain of negative charge by a substance, often accompanied by loss of oxygen or the addition of hydrogen.

Rhizobia (*rhizo* = root; *bios* = life): The group of bacteria in the genus *Rhizobium* that live symbiotically in root nodules of legumes. In the root nodules, the rhizobia obtain energy from the plants to convert dinitrogen gas in the air to compounds of nitrogen that the plants can use.

Rhizomorph: An aggregation of many parallel filaments (hyphae) of fungi that grows as a single unit.

Rhizosphere: The region of soil immediately surrounding plant roots. Bacteria and other microbes seem to be especially abundant in this zone.

Saprophytes or **Saprobes** (*sapros* = rotten; *bios* = life): Organisms that feed on nonliving organic matter. They include detritivores, decomposers, and scavengers.

Silt: A class of inorganic soil particles measuring between 0.002 and 0.05 mm in diameter. Silt particles are silky to the touch and arise from the weathering of rocks.

Soil horizon: A layer of soil that usually lies parallel to the surface of the ground. Each layer can have characteristic chemical and physical properties.

Soil structure: The arrangement of soil particles into naturally occurring clumps called aggregates. If the clumping of soil particles is caused by tillage, the clumps are called clods.

Soil texture: The relative proportion of sand, silt, and clay in a soil.

Subsoil: The layer of soil (horizon) that lies below the topsoil; the layer that is not turned during cultivation.

Symbiosis: An intimate, continuous, and mutually beneficial interaction between two different organisms.

Tillage: The mechanical disruption of soil that modifies soil conditions for production of crops.

Topsoil: The layer of the soil that is disrupted and mixed during tillage. This layer is also known as the A horizon.

Urogomphi: A pair of dorsal processes that project from the posterior end of the ninth segment of a beetle larva's abdomen.

Vertebrates (*vertebra* = segment): Multicellular animals with backbones. The backbones consist of segments called vertebrae.

Vesicular-arbuscular mycorrhizae: *See* Mycorrhizae.

Zoospore (*zoon* = animal; *spora* = seed): A swimming spore that is produced asexually.

Further Reading

GENERAL

Baskin, Yvonne. *Under Ground: How Creatures of Mud and Dirt Shape our World.* Washington: Island Press, 2005.

Bial, Raymond. *A Handful of Dirt.* New York: Walker, 2000.

Hillel, David J. *Out of the Earth: Civilization and the Life of the Soil.* New York: Free Press, 1991.

Ingham, Elaine, R., Andrew R. Moldenke, and Clive Edwards. *Soil Biology Primer.* Washington: United States Department of Agriculture, Natural Resources Conservation Service, 1999.

Logan, William B. *Dirt: The Ecstatic Skin of the Earth.* New York: Riverhead, 1995.

Martin, Deborah L., and Karen Costello Soltys, eds. *Soil: Rodale Organic Gardening Basics.* Emmaus, PA: Rodale, 2000.

Wolf, David W. *Tales from the Underground: A Natural History of Subterranean Life.* Cambridge, MA: Perseus, 2001.

GENERAL, TECHNICAL

Adl, Sina. *The Ecology of Soil Decomposition.* Wallingford, UK: CABI Publishing, 2003.

Bardgett, Richard. *The Biology of Soil: A Community and Ecosystem Approach.* Oxford: Oxford University Press, 2005.

Brady, N. C., and R. R. Weil. *Elements of the Nature and Properties of Soils.* Upper Saddle River, NJ: Prentice Hall, 2000.

Coleman, David C., D. A. Crossley, and Paul F. Hendrix. *Fundamentals of Soil Ecology.* Amsterdam: Elsevier, 2004.

Dindal, Daniel L., ed. *Soil Biology Guide.* New York: Wiley, 1990.

Killham, Ken. *Soil Ecology.* Cambridge: Cambridge University Press, 1994.

MICROBES

Dusenberry, David B. *Life at Small Scale: The Behavior of Microbes.* New York: Scientific American Library, 1996.

Dyer, Betsey D. *A Field Guide to Bacteria.* Ithaca, NY: Cornell University Press, 2003.

FUNGI AND NONFLOWERING PLANTS

Hudler, George W. *Magical Mushrooms, Mischievous Molds.* Princeton, NJ: Princeton University Press, 1998.

Lincoff, Gary H. *National Audubon Society Field Guide to North American Mushrooms.* New York: Knopf, 1981.

Shuttleworth, Floyd S., and Herbert S. Zim. *Mushrooms and Other Non-flowering Plants.* New York: Golden, 1987.

INVERTEBRATES

Borror, Donald J., and Richard E. White. *A Field Guide to the Insects of America North of Mexico.* Boston: Houghton Mifflin, 1998.

Buchsbaum, Ralph M., J. Pearse, and V. Pearse. *Animals without Backbones: An Introduction to the Invertebrates.* Chicago: University of Chicago Press, 1987.

Chu, Hung-fu. *How to Know the Immature Insects.* Dubuque: William C. Brown, 1992.

Levi, Herbert W., and Lorna R. Levi. *Spiders and Their Kin.* Revised ed. New York: Golden, 1990.

Stewart, Amy. *The Earth Moved: On the Remarkable Achievements of Earthworms.* Chapel Hill, N.C.: Algonquin, 2004.

VERTEBRATES

Halliday, Tim, and Kraig Adler, eds. *The New Encyclopedia of Reptiles and Amphibians.* Oxford: Oxford University Press, 2002.

Macdonald, David, ed. *The New Encyclopedia of Mammals.* Oxford: Oxford University Press, 2001.

Perrins, Christopher, and C. J. O. Harrison. *Birds: Their Life, Their Ways, Their World.* Pleasantville, NY: Reader's Digest Association, 1979.

Tyning, Thomas F. *A Guide to Amphibians and Reptiles.* Boston: Little, Brown and Company, 1990.

Zim, Herbert S., and Donald F. Hoffmeister. *Mammals: A Guide to Familiar American Species.* New York: Golden, 1991.

ORGANIC GARDENING AND COMPOSTING

Appelhof, Mary. *Worms Eat My Garbage.* Kalamazoo, MI: Flowerfield, 1997.

Campbell, Stu. *Let It Rot! The Gardener's Guide to Composting.* Pownal, VT: Storey, 1998.

Lowenfels, Jeff, and Wayne Lewis. *Teaming with Microbes: A Gardener's Guide to the Soil Food Web.* Portland, OR: Timber Press, 2006.

Martin, Deborah L., and Grace Gershuny, eds. *The Rodale Book of Composting.* Emmaus, PA: Rodale, 1992.

Nancarrow, Loren, and Janet Hogan Taylor. *The Worm Book.* Berkeley: Ten Speed Press, 1998.

Smillie, Joe, and Grace Gershuny. *The Soul of Soil: A Soil-Building Guide for Master Gardeners and Farmers.* White River Junction, VT: Chelsea Green, 1996.

Index

Note: Page numbers in italics refer to figures.